AIRO Springer Series

Volume 14

The **AIRO Springer Series** focuses on the relevance of operations research (OR) in the scientific world and in real life applications.

The series publishes peer-reviewed only works, such as contributed volumes, lectures notes, and monographs in English language resulting from workshops, conferences, courses, schools, seminars, and research activities carried out by AIRO, Associazione Italiana di Ricerca Operativa – Optimization and Decision Sciences: http://www.airo.org/index.php/it/.

The books in the series will discuss recent results and analyze new trends focusing on the following areas: Optimization and Operation Research, including Continuous, Discrete and Network Optimization, and related industrial and territorial applications. Interdisciplinary contributions, showing a fruitful collaboration of scientists with researchers from other fields to address complex applications, are welcome.

The series is aimed at providing useful reference material to students, academic and industrial researchers at an international level.

Should an author wish to submit a manuscript, please note that this can be done by directly contacting the series Editorial Board, which is in charge of the peer-review process.

THE SERIES IS INDEXED IN SCOPUS

Michele Barbato · Annarita De Maio ·
Giusy Macrina · Lorenzo Peirano · Marco Premoli
Editors

Beyond Frontiers of Operations Research: Emerging Technologies and Innovative Optimization Paradigms

Proceedings of the 7th AIROYoung Workshop (February 15–17, 2023, University of Milan, Italy) and 8th AIROYoung Workshop (February 14–16, 2024, University of Calabria, Italy)

AIRO
ASSOCIAZIONE ITALIANA DI RICERCA OPERATIVA
OPTIMIZATION AND DECISION SCIENCE

Springer

Editors
Michele Barbato
Department of Computer Science
University of Milan
Milan, Italy

Annarita De Maio
Dipartimento di Economia,
Statistica e Finanza "G. Anania"
Università della Calabria
Rende, Italy

Giusy Macrina
Dipartimento di Ingegneria Meccanica,
Energetica e Gestionale
Università della Calabria
Rende, Italy

Lorenzo Peirano
Department of Economics and Management
University of Brescia
Brescia, Italy

Marco Premoli
Department of Computer Science
University of Milan
Milan, Italy

ISSN 2523-7047 ISSN 2523-7055 (electronic)
AIRO Springer Series
ISBN 978-3-031-85893-2 ISBN 978-3-031-85894-9 (eBook)
https://doi.org/10.1007/978-3-031-85894-9

This Springer imprint is published by the registered company Springer Nature Switzerland AG
The registered company address is: Gewerbestrasse 11, 6330 Cham, Switzerland

Preface

The AIRO Young Researchers Chapter (AY for short) is part of the Italian Operational Research Society (AIRO). Through networking and resource-sharing initiatives, AY aims to promote collaboration and career advancement opportunities for students and early-career researchers with an interest in Operations Research.

Every year, AY organizes the AIROYoung workshop at an Italian university. Since its first edition in 2017, the AIROYoung Workshop has gained notable recognition, attracting an increasing number of participants from both academia and industry and becoming a reference for the young operational researchers and practitioners in Italy and abroad.

This volume contains nine accepted contributions to the proceedings of the 7th and 8th AIROYoung workshops. The 7th AIROYoung workshop, entitled "Operations Research Beyond Frontiers", was held at the University of Milan, from 15th to 17th February 2023. The aim of the workshop was to bring together young operational researchers working on innovative applications and methods of Operations Research, with an emphasis on applications to computer science and statistical data analysis. The 8th AIROYoung workshop, entitled "Emerging Technologies for Decision Support Systems and Innovative Optimization Paradigms", was held at the University of Calabria, from 14th to 16th February 2024. The focus was on innovative optimization paradigms and applications of emerging technologies in decision support systems.

Over half of the contributions to this volume are research works that focus on the interaction between Operations Research and Machine Learning.

The paper "From Roads to Rails: Bridging the Gap in Signal Recognition Through Computer Vision" by Pappaterra and Flammini provides a comparative literature review of traffic signal recognition via Computer Vision. Moreover, the paper explores potential applications of this methodology to decision making within the railway domain.

The paper "On the Hardness of Edge Coloring Problems Associated with Scheduling" by Barbato, Delle Donne and Lancini, delves into the complexity of edge-coloring problems arising in scheduling and federated vertical learning, proving them to be NP-hard.

A more specific application of Machine Learning and Operations Research interactions is presented in the paper "Machine Learning Algorithms for Supporting Optimization in Defining Yard Stacking Strategies" by Ambrosino and Xie. The authors propose an unsupervised learning approach to cluster similar containers present at a maritime terminal, and define a mixed-integer linear program to organize the clustered containers in the storage yard.

The paper proposed by Ferrisi, entitled "Assessing the Academic Performance of Italian Students Using an Explainable Machine Learning Approach" introduces a novel student performance indicator for the identification of at-risk students via machine learning models. Then, the author applies explainable Artificial Intelligence techniques to interpret the models' predictions.

In their paper, "Integrating Human-Centric AI in Fashion Industry Optimization: A Step Towards Industry 5.0" Dragone, Battaglia and Cerrone present an application of mathematical programming and machine learning to the high-fashion clothing industry. Their approach exploits large language models to translate operators' feedback expressed in natural language into formal objectives of a linear programming model for optimizing fabric cutting operations.

The remaining contributions present applications of mathematical programming techniques to contemporary optimization challenges.

The paper "Column Generation Algorithms for the Pickup and Delivery Problem with Time Windows and Last-in-First-out Loading" by Palasgo, Fischetti and Roberti studies a logistics problem with the goal of minimizing the travel cost to serve a set of customers while respecting the customers' time window constraints and the last-in-first-out policy of the pickup and delivery operations. The study proposes a lower bound procedure for the problem based on a column generation approach. The experimental results show that the obtained bounds are tight and that the algorithmic procedure is effective.

The paper "Meal Delivery with Unmanned Aerial Vehicle: A Case Study from the University of Calabria" by Ciacco, Luzzi and Scalzo investigates the potential usage of drones for the delivery of food from canteens to students on the campus of the University of Calabria. A bi-objective program is proposed whose goal is to minimize both the distance traveled and the travel time. The bi-objective program is solved using an epsilon-constraint approach. The obtained solutions show that the usage of unmanned aerial vehicles is viable for this particular application.

In the paper "An Effective Matheuristic Approach to Solve Nurse Rostering Problem", Belfiore proposes a matheuristic approach to a challenging nurse rostering problem. In the proposed framework, nurses are assigned to shifts according to their skills, day preferences, contract requirements and hospital needs. The proposed matheuristic is composed of two steps: first it derives a feasible solution from the optimal solution of a problem relaxation; then, it combines a fix-and-optimize approach with a large neighborhood search.

In conclusion, the paper by Maragò, entitled "Mixed-Integer Linear Programming Models for Water-Energy-Food-Land Nexus Optimization with Land Degradation Phenomenon" deals with the optimization of a food production system under the Water-Electricity-Food-Land Nexus constraints. These latter impose the satisfaction of electricity and food demands in a region while limiting greenhouse gas emissions and land degradation, and without exceeding the availability of fossil fuels and water used to produce energy and food. The paper presents two mixed-integer linear programs that allow to decide which power plants to activate and expand. A what-if analysis is conducted by solving the proposed optimization programs on realistic instances.

As editors of the volume, we would like to thank AIRO and AY for their financial and organizational support to the 7th and 8th AIROYoung workshops. Furthermore, we extend our gratitude to the Department of Computer Science of University of Milan, the Department of Mechanical, Energy and Management Engineering and the Department of Economics, Statistics and Finance of the University of Calabria for hosting the workshops. We would like to thank the invited lecturers, the industrial representatives and sponsors and the authors who contributed their talks and their articles to the workshops and proceedings. We would also like to thank the researchers who participated in the review process of this volume. Finally, we would like to express our gratitude to Springer for its strong support and cooperation during the event and the publishing process.

Milan, Italy — Michele Barbato
Rende, Italy — Annarita De Maio
Rende, Italy — Giusy Macrina
Brescia, Italy — Lorenzo Peirano
Milan, Italy — Marco Premoli

Contents

Interactions Between Operations Research and Machine Learning

From Roads to Rails: Bridging the Gap in Signal Recognition Through Computer Vision

Mauro José Pappaterra and Francesco Flammini

Abstract In this paper, we present a comprehensive review of traffic signal recognition through Computer Vision (CV) and its potential applications for operations and decision making within the railway domain. This methodology is well-established in the automotive industry (i.e., self-driving cars) where several specific methods have been developed, but little is written on signal recognition for autonomous trains or as Advanced Driver Assistance Systems (ADAS) for railways. While onboard Automatic Train Protection (ATP) systems currently dominate signal analysis in railways, CV-based traffic sign detection and recognition systems have significant potential for the industry. The objective of this study is to examine the state-of-the-art CV techniques applied to road signals, the challenges, and possible solutions for accurate and efficient signal detection and classification; and how these techniques can be incorporated into the railway sector. We also conduct a short literature review of traffic sign detection and recognition in both roads and railway systems. This study discusses CV systems utilized in road signals in terms of their suitability for railway applications and explores the advantages and challenges of utilizing these existing technologies for signal recognition in railway environments.

Keywords Railways · Autonomous trains · Computer vision · Signal recognition · Artificial intelligence · Machine learning · Driver assistance · ADAS

M. J. Pappaterra (✉)
Uppsala University, Uppsala, Sweden
e-mail: mauro.pappaterra@ieee.org; mauro.pappaterra@gmail.com

F. Flammini
University of Applied Sciences and Arts of Southern Switzerland—Dalle Molle Institute for Artificial Intelligence (IDSIA USI-SUPSI), Lugano, Switzerland
e-mail: francesco.flammini@ieee.org

M. Barbato et al. (eds.), *Beyond Frontiers of Operations Research: Emerging Technologies and Innovative Optimization Paradigms*, AIRO Springer Series 14,
https://doi.org/10.1007/978-3-031-85894-9_1

1 Introduction

This paper studies the potential application of Artificial Intelligence (AI) for Traffic Signs (TS) and Traffic Lights (TL) detection and recognition in the railway industry. Research in Traffic Sign Detection and Recognition (TSDR) is well established in the automotive industry [1], but little is written on traffic signals for railway operations [2, 3]. Automated TSDR systems have different applications in operational research including maintenance of highways and city roads, sign inventory keeping, and of course applications for both Advanced Driver Assistance Systems (ADAS) and intelligent autonomous vehicles [4]. When it comes to modern railways, the status of the signalling systems is transmitted directly to the train onboard Automatic Train Protection (ATP) system. Consequently, Computer Vision (CV) models are not the primary focus of study for signal analysis. Nonetheless, there are some situations where AI-based TSDR systems could be implemented in railways. For instance, in the case of developing countries, where no modern signaling and ATP systems are present in the entirety of the national railway network. In the case of countries that do count modern infrastructure, intelligent TSDR systems can be used when there is a technical failure in the ATP system [5]. The implementation of this technology is relevant to Operational Research (OR) for several reasons including optimization of resources, enhancement of safety, enabling data-driven decision making, and improvement in cost effectiveness, as well as on the reliability and redundancy of the railway system. Moreover, the continuous monitoring and analysis of TS and TL signals contributes to the consistent and reliable operation of railway services.

The purpose of this study is two-fold; to make a quick survey of the state-of-the-art CV applications for TS and TL detection and recognition, and to analyse its possible applications in the railway sector. For this purpose, we first analyse existing literature reviews and surveys, along with some newer publications in the matter; we are particularly interested in real-time detection and classification of both TS and TL in mainline trains using common CV techniques, most specifically Artificial Neural Networks (ANN), Deep Learning (DL) and Convolutional Neural Networks (CNN), with an emphasis on ADAS and solutions related to autonomous vehicles.

2 Literature and Open Datasets Related to TSDR in Railways

A small number of studies have been published on the topic of TSDR in railways. We have identified 11 relevant studies in total; This is much smaller in comparison with the literature related to TSDR in automobiles. The study [6] proposed a CV-based solution for the detection and recognition of Italian railway signals. The authors in [7] present a backpropagation ANN model for the detection and recognition of a particular class of Polish railway signs. While the authors in [8] implement an inception method for the detection of the railway track and a Faster Region-Based Convolutional Neural Network (R-CNN) for the detection of railway signals. The study [9]

Table 1 List of public datasets, including some technical details

Dataset	No. images	No. classes	Image resolutions	Country	Year
GERALD	5.000	59	1920 × 1080 1280 × 720	Germany	2023
FRSign	105.352	19	2048 × 1536 1920 × 1200	France	2020

applies Deep Neural Networks, more precisely the You Only Look Once (YOLO) v3 CNN for the detection and recognition of railway traffic signals in real-time. The study [10] evaluates the security and safety implications for a CV-based system for the detection and recognition of railway signals. The authors in [11] present a hybrid solution for TSDR that combines classic image processing algorithms, a Haar classifier and a YOLO single-shot detector (SDD) network. While the authors in [12] proposed a CV model based on the OpenCV library to recognize railway signs and tracks, as well as detecting railway workers. The authors in [13] present a real-time detection method to recognize locomotive signal lights and railway pedestrians based on an improved YOLOv4. Finally, the study presented in [14] implements DL for the detection and recognition of railway traffic signs. This solution is based in YOLOv5 and a sequence-based blink detection heuristic model for the detection of LS states.

At the time of writing, there are only 2 public datasets that are aimed to TSDR in railways: the French Railway Signalling (FRSign) dataset [15]; and the German Railway Lightsignal Dataset (GERALD) [16] (Table 1). These datasets contain railway scene images annotated with bounding boxes for signals, and context such as weather and light conditions (GERALD) and signal type and state (FRSign). Other open datasets that do include images of railway signals—though not as the main focus—are RailSem19, CityScapes and the Common Objects in Context (COCO) datasets [17, 18].

3 Existing Literature Reviews on Road TSDR

For this study we revised existing literature reviews and surveys in the different applications of AI for TS and TL detection and recognition for road operations.

The earliest review [4] published in 2002 presents a survey on TSDR for roads, presenting around 30 papers written in the subject, divided in two categories, signs detection (subclassified in segmentation through colour thresholding and border detection) and sign recognition. Another study [19]—published a few years later—implemented a similar classification. Both papers point to ANNs as the most suitable alternative for the task of TSDR at the time of writing.

In 2010 the study [20] presented an exhausting quantitative comparison of segmentation methods used for TSR including colour-space thresholding, edge detection, and chromatic/achromatic decomposition. In this decade, many new datasets aimed at TSDR applications were presented [21–23]. This allowed for better surveys and benchmarking of different methods when writing reviews. For instance, the study

[24] introduced a survey of different Hough-like methodologies for the detection of traffic signals and different preprocessing steps using the Stereopolis dataset [25] for benchmarking. The authors in [26] compare the performance of human recognition and a baseline linear discriminant analysis (LDA) with the top performing solutions for TSR presented for the competition using the German Traffic Sign Recognition Benchmark (GTSRB) for comparison [21]. Later, the authors in [22] presented the results of a competition for traffic sign detection along with the German Traffic Sign Detection Benchmark (GTSDB) dataset, which was implemented for the comparison of the models proposed by the contestants in said vying. Though the GTSRB and GTSDB datasets are centered in traffic signs and lights found in European roads (and other countries that adhere to the Vienna Convention on Road Signs and Signals); These datasets will become very popular among researchers and will be used to benchmark different models. Lastly, the authors in [23] present a literature review of 41 different vision-based solutions for the detection of traffic signs. The study makes emphasis is in the signal detection stage and in human-centered ADAS systems. The authors present the University of California's Laboratory for Safe and Intelligent Automobiles (LISA) dataset. By the end of 2013, there were at least 8 different datasets publicly available for TSDR research, most of which were still centered in European signals.

For the upcoming years, there will be a surge in research related to TSDR, based on many factors: the hype for autonomous driving vehicles, the availability of research datasets, and the breakthrough in CV technology. In 2021, the study [27] provided with an updated overview of at least 18 different datasets in current time created for the purpose of benchmarking and evaluating TSDR solutions. Authors in [28] benchmark different Deep Convolutional Neural Networks (DCNNs) and their architectures for TSR. This study implements the GTSRB dataset for comparison purposes. The most recent study [29] published in 2022 presents and compares 6 different CNNs for TSR based on their predictions using both the GTSRB and the LISA datasets.

4 Latest Developments of TSDR in Roads

For the purpose of this study, we have analysed the latest developments found in literature. It shows new CV approaches to TSDR in roads. For instance, the study [30] proposes a hierarchical approach based on shape detection and image classification. This solution implements the Mask R-CNN deep learning model to detect traffic signs based on their shapes and the Xception model for their classification. The authors in [31] present an end-to-end CNN for that can simultaneously detect and classify traffic signs. This solution leverages on the fact the traffic signs usually occupy only a small proportion of the input images. The authors of this study introduce the dataset Tsinghua-Tencent 100 K, which is used as a benchmarking parameter. This dataset is used in the study [32] which introduces a Multi-scale R-CNN for TS detection and classification. For the detection stage, the MR-CNN implements multiscale deconvolution while multi-scale contextual regions are implemented for the classification stage.

Authors in [33] evaluate the use of Vision Transformers for TSR. The study includes results of experiments with seven CNNs and five Vision Transformer models, applied for the recognition of signals in open datasets. The CNNs include VGG16, ResNet, DenseNet, MobileNet, SqueezeNet, ShuffleNet, and MnasNet. Authors in [34] proposed an improved Sparse R-CNN, an ANN model inspired on the DL Transformer architecture. And the study [35] presents an optimized Region Proposal Network (RPN) that implements weighted loss and hard-soft distribution-wise distillation loss for cutting channels. This study dwells in the possibility of using knowledge distillation for TSR. The model uses VGG16 network and Faster R-CNN Network to accomplish this. This solution can be used to support the transfer of learning between different size CNNs. While the authors in [36] implement an improved YOLOv4 algorithm for the detection and recognition of traffic lights. To overcome the original algorithm's insensitivity to small objects and low detection precision in traffic light detection and recognition.

Different smart solutions are present for the reduction and optimization of parameters in CV algorithms. The authors in [37] propose a small network MicronNet-BF for TS classification. This small but powerful CNN is improved by batch normalization and factorization that reduces parameters and improves accuracy. While authors in [38] present a fully integer-based CNN accelerator on Field Programmable Gate Arrays (FPGA) for real-time TS recognition. This low-cost solution is aimed for low-resources and computational power networks, by constructing the CNN so all operations can use only integer arithmetic. Another example [39], proposes a real-time ADAS prototype that includes a model for TS and road objects detection based on a grid partition technique. The proposed model implements the YOLOv5 algorithm, and it is optimized using mobile GPU platforms. Moreover, the authors in [40] propose a new approach for improving the speed and accuracy of TSDR, based on focusing on different Regions of Interest (ROIs) in images and parallelization. The authors present a short survey on different YOLO algorithms used for TSDR.

New solutions based on CV also address some of the recurrent challenges in TSDR, such as occlusion of signals, weather conditions and other obstacles. For example, the authors in [41] present a cognition-based TS recognition model for partially occluded traffic signs based on CNN. The model focuses on visible parts of TS, and suppresses occluded regions, by implementing a CNN-based invariant feature extraction technique that is augmented with feature interaction dimensionality reduction. The authors in [27] proposed two CV models for real-time TSDR and fog detection for ADAS systems. For TSD the model proposed implements ACF detector while the TSR stage is solved with a CNN. For fog detection the authors implement an image feature extraction combined with RGB, HSV and XYZ colour spaces. The study [42] proposes three different Deep Neural Networks (DNNs) for TSR under the possibility of camera failures, including blurred images, high brightness, broken lenses, chromatic aberrations, condensation, dead pixels, dirt, ice, and rain among others. To do this the study evaluates single frame and sliding window classifiers. The authors in [43], propose a preprocessing statistical boosting layer to facilitate TSR of occluded and degraded TS using feature extraction or CNN-based techniques. This solution enhances the visible portions of the sign, while suppressing the occluded parts which facilitates extracting/learning invariant features. Finally, the authors in [44] proposed a DL solution for the recognition of traffic lights in varying illumination scenarios. This solution implements a CNN to extract and detect features from camera images and implements a saliency map containing the traffic light location.

5 Physicality of Railway Signals

To assess how AI techniques for road TSDR can be applied to the railway industry, it is important to make parallels between the physicality of road and railway traffic signs. Railway signs can be divided into four big categories: semaphore, colour light, ground-mounted and trackside signals (Fig. 1). Semaphore signals count with a blade that can be pivoted to different angles and a spectacle that holds colour lenses that move in front of a lamp to provide different indications. Colour light signals are electrically controlled lamps and coloured lenses enclosed in a modular unit. Lights can be combined in different arrangements and displays (solid or flashing lights) to convey different cautionary aspects and indications. Some signals have feathers that count with smaller sets of lights. Then we have ground-mounted signals (also known as dwarf signals), which are mini disk semaphores or colour lights used for indications related to shunting manoeuvres.

Similar to their regular-size counterparts, the signal lights can be arranged in different configurations and states to convey different messages. Finally, trackside signs are more conventional signposts which are typically made of metal or fibre, and the most similar to classic road signals. They are mostly implemented for speed limits and stop signals used to protect junctions, level crossings and block sections. Other trackside signals can also be used to indicate the beginning of neutral sections, whistle boards, station limits, radio frequency boards and more.

Other signals found in railways include information signs, platform numbers and letters, signal repeaters, fixed distant signal boards, station limit boards and more.

6 Discussion on TSDR Transferability from Roads to Railways

In this study we compared papers related to TSDR applied to railways and roads. There is a big disparity in the number of papers found among each of the transport areas, in which we have observed that most of the literature is centered on road signals. We also consulted literature reviews on TSDR applied to roads; but with the exception of [22, 29] all of these reviews and surveys are more than a decade old. Researchers agreed that that the lack of public datasets was one of the factors for the lack of reviews in the subject [21, 23]. The reviews in TSDR in roads notably increased

Fig. 1 Different examples of colour light signals, semaphore signals, trackside signs and ground-mounted signals. All images taken from Rail Sim World 3; a simulator developed by Dovetail Games

along with the publications of open datasets in the 2010 and more recently. Another factor that hinders the research in the topic, is the lack of uniformity among papers: the stages of traffic detection and recognition are not always present in all models, some studies center in the former [22, 23, 25] and some in the latter [26, 29], and some studies present unified solutions [6, 19, 40] that include both signal detection and recognition. When it comes to railway signals, we were able to identify 11 papers and only 2 datasets directly focused in TSDR (plus 3 other datasets that could be circumvented for related research). On the other hand, there are at least 20 datasets related to TSDR on roads, which might explain the disparity in research in the topic between roads and railways.

When it comes to modern railways, the status of the signalling systems is transmitted directly to the train onboard ATP system, which can automatically enforce speed restrictions, apply brakes, and prevent collisions based on the received signals. Consequently, CV models are not the primary focus of study for signal analysis in railway transport. However, we have identified specific scenarios where CV-based TSDR systems could still be highly beneficial. These situations can be broadly categorized into two main contexts: developing countries (with limited railway infrastructure) and developed countries (with modern infrastructure). Developing countries might have extensive railway networks (e.g. Indian Railways large national network) but may lack the financial resources to upgrade to modern signalling and ATP systems. Quite often, in such environments, traditional manual signaling methods are still in use. In this context, CV-based TSDR systems can provide a cost-effective solution to enhance safety and efficiency in railway operations. When it comes to developed countries with modern railway infrastructure, a secondary CV-based TSDR system can serve as a crucial backup in the eventual case of a technical failure in the ATP system, enhancing the safety and redundancy of the network. CV-based systems can also enhance operational continuity, by minimizing downtime and service disruptions caused by ATP system failures. Beyond these two contexts mentioned above, intelligent TSDR systems can also contribute to other aspects of railway operations, such as predictive maintenance, data analytics and optimization, and integration with other smart transportation systems to create a more cohesive and intelligent transportation network.

To assess how the latest AI techniques for TSDR in roads can be transferred to railways, it is important to make parallels between road and railway traffic signs. Railway signs can be divided into four big categories: semaphore, colour light, ground-mounted and trackside signals. The latter are the most similar to road signals, while the rest can be equiparated to road traffic lights and semaphores. But unlike road traffic signals, there are no conventions among countries regarding railway signalling. Because of this, the convention and the level of complexity of railway signals can vary greatly in different areas. Railway signs are quite complex in comparison, different types of signals count with different parts and light systems; These can be arranged in different patterns to display different messages or convey different levels of precaution. Nonetheless, we believe that this issue can be easily addressed by adjusting the AI-based TSDR system based on the train location, and adjusting the AI model to the particularities of the signals of the network.

When it comes to ADAS; contrary to automobile systems—which are quite flexible—trains are much more conservative systems. ADAS will not be accepted unless they are demonstrated to be safe and reliable according to national standards and regulations. Our research shows that AI-based TSDR systems can detect and interpret traditional signals, sending real-time information to train operators or centralized control systems. This can significantly improve the safety of railway operations, by reducing human error, which is a common cause of accidents in manually operated systems. This is another reason why more efforts in the scope of the study of CV-based TSDR in railways are necessary.

7 Conclusion

In the past years there has been an increase in research for Computer Vision (CV) methods for traffic sign detection and recognition (TSDR) on roads. This is a product of recent developments in Artificial Intelligence (AI), mostly Convolutional Neural

Networks (CNN), and Transformers and the availability of public datasets for the study of TSDR in roads. A lot of these advances and research for roads traffic signs could be directly transferred to railways. Nonetheless, there are some challenges: such as the complexity of railway signals and the lack of unified systems from country to country. More rigorous safety controls on trains ADAS are also another factor to consider. Moreover, we believe that the lack of openly available datasets for railway traffic signs is another big factor in the lack of research in the area.

While modern ATP systems currently dominate signal analysis in developed railway networks, CV-based TSDR systems have significant potential in various contexts. For instance, deploying CV and AI systems is relatively less expensive compared to overhauling the entire railway signaling infrastructure. This makes it a viable option for developing countries looking to incrementally improve their rail safety and operations. Even in countries with advanced railway systems, technical failures in ATP systems can occur. An intelligent TSDR system can act as a failsafe mechanism, ensuring continuous monitoring and interpretation of signals even when the primary ATP system is compromised. We emphasize the importance of automated TSDR for ensuring the safety, efficiency, and reliability of semi-supervised and autonomous railway operations: hoping that this short study will foster research on traffic signs detection and recognition in railways. The next step is to analyse the key components of CV systems utilized for automobile TSDR in terms of their suitability for railway signals; and explore the advantages and challenges of utilizing these existing technologies for signal recognition in railway environments. The application of AI in railway signaling is a promising avenue for the future of rail transportation. By making more exhaustive research and identifying the points in common, the railway industry can benefit from the advances in traffic sign detection and recognition done in the automotive industry.

Acknowledgements The work of the authors was partly supported by the Swiss State Secretariat for Education, Research and Innovation (SERI) under contract no. 24.00528 (PhDs EU-Rail project). The project has been selected within the European Union's Horizon Europe research and innovation programme under grant agreement no. 101175856. Views and opinions expressed are however those of the authors only and do not necessarily reflect those of the funding agencies, which cannot be held responsible for them.

References

1. European Parliamentary Research Service (EPRS), European Parliament. Artificial Intelligence in Transport: Current and Future Developments, Opportunities and Challenges, April 2019. Available online: 9/635609/EPRSBRI(2019)635609_EN.pdf. Accessed 14 May 2023
2. Bešinović, N., Tang, R., Lin, Z., Liu, R., Tang, T., De Donato, L., Vittorini, V., Wang, Z., Flammini, F., Pappaterra, M.J., et al.: Deliverable D1.2 summary of existing relevant projects and state-of-the-art of AI application in railways. RAILS, Shift2Rail (2021). Available online: 12_v23.pdf. Accessed 15 May 2023
3. Tang, R., De Donato, L., Bešinović, N., Flammini, F., Goverde, R.M.P., Lin, Z., Liu, R., Tang, T., Vittorini, V., Wang, Z.: A literature review of artificial intelligence applications in railway

systems. Transp. Res. Part C: Emerg. Technol. **140**, 103679 (2022). ISSN 0968-090X. https://doi.org/10.1016/j.trc.2022.103679
4. de la Escalera, A., Armingol, J.M., Mata, M.: Traffic sign recognition and analysis for intelligent vehicles. Image Vis. Comput. **21**(3), 247–258 (2003). ISSN 0262-8856. https://doi.org/10.1016/S0262-8856(02)00156-7
5. Flammini, F., De Donato, L., Fantechi, A., Vittorini, V.: A vision of intelligent train control. In: Collart-Dutilleul, S., Haxthausen, A.E., Lecomte, T. (eds.) Reliability, Safety, and Security of Railway Systems. Modelling, Analysis, Verification, and Certification. RSSRail 2022. Lecture Notes in Computer Science, vol. 13294. Springer, Cham (2022). https://doi.org/10.1007/978-3-031-05814-1_14
6. Marmo, R., Lombardi, L., Gagliardi, N.: Railway sign detection and classification. In: IEEE Intelligent Transportation Systems Conference, Toronto, ON, Canada, vol. 2006, pp. 1358–1363 (2006). https://doi.org/10.1109/ITSC.2006.1707412
7. Mikrut, S., Mikrut, Z., Moskal, A., Pastucha, E.: Detection and recognition of selected class railway signs. Image Process. Commun. **19**(2–3), 83–96 (2015). https://doi.org/10.1515/ipc-2015-0013
8. Ritika, S., Mittal, S., Rao, D.: Railway track specific traffic signal selection using deep learning (2017). arXiv preprint arXiv:1712.06107
9. Etxeberria-Garcia, M., Ezaguirre, F., Plazaola, J., Muñoz, U., Zamalloa, M.: Embedded object detection applying deep neural networks in railway domain. In: 2020 23rd Euromicro Conference on Digital System Design (DSD), pp. 565–569 (2020). https://doi.org/10.1109/DSD51259.2020.00093
10. Boulineau, J.F.: Safe recognition A.I. of a railway signal by on-board camera. In: Bernardi, S., et al. (eds.) Dependable Computing—EDCC 2020 Workshops. EDCC 2020. Communications in Computer and Information Science, vol. 1279. Springer, Cham (2020). https://doi.org/10.1007/978-3-030-58462-7_1
11. Chołodowicz, E., Lisiecki, P., Lech, P.: Hybrid algorithm for the detection and recognition of railway signs. In: Burduk, R., Kurzynski, M., Wozniak, M. (eds.) Progress in Computer Recognition Systems. CORES 2019. Advances in Intelligent Systems and Computing, vol. 977. Springer, Cham (2020). https://doi.org/10.1007/978-3-030-19738-4_34
12. Shin, D., Jin, J., Kim, J.: Enhancing railway maintenance safety using open-source computer vision. J. Adv. Transp. **2021**, 1–8 (2021)
13. Wang, H., Pei, H., Zhang, J.: Detection of locomotive signal lights and pedestrians on railway tracks using improved yolov4. IEEE Access **10**, 15495–15505 (2022). https://doi.org/10.1109/ACCESS.2022.3148182
14. Staino, A., Suwalka, A., Mitra, P., et al.: Real-time detection and recognition of railway traffic signals using deep learning. J. Big Data Anal. Transp. **4**, 57–71 (2022). https://doi.org/10.1007/s42421-022-00054-7
15. Harb, J., Rébéna, N., Chosidow, R., Roblin, G., Potarusov, R., Hajri, H.: FRSign: a large-scale traffic light dataset for autonomous trains (2020). arXiv preprint arXiv:2002.05665
16. Leibner, P., Hampel, F., Schindler, C.: GERALD: a novel dataset for the detection of German mainline railway signals. Proc. Inst. Mech. Eng., Part F: J. Rail Rapid Transit. **237**(10) (2023). https://doi.org/10.1177/09544097231166472
17. Pappaterra, M.J., Flammini, F., Vittorini, V., Bešinović, N.: A systematic review of artificial intelligence public datasets for railway applications. Infrastructures **6**, 136 (2021). https://doi.org/10.3390/infrastructures6100136
18. Lin, TY., et al.: Microsoft COCO: common objects in context. In: Fleet, D., Pajdla, T., Schiele, B., Tuytelaars, T. (eds.) Computer Vision—ECCV 2014. ECCV 2014. Lecture Notes in Computer Science, vol. 8693. Springer, Cham (2014). https://doi.org/10.1007/978-3-319-10602-1_48
19. Fleyeh, H., Dougherty, M.: Road and traffic sign detection and recognition. In: 10th EWGT Meeting and 16th Mini-EURO Conference, Poznan, Poland, 13–16 September 2005 (2005)
20. Gomez-Moreno, H., Maldonado-Bascon, S., Gil-Jimenez, P., Lafuente-Arroyo, S.: Goal evaluation of segmentation algorithms for traffic sign recognition. IEEE Trans. Intell. Transp. Syst. **11**(4), 917–930 (2010). https://doi.org/10.1109/TITS.2010.2054084. Dec

21. Stallkamp, J., Schlipsing, M., Salmen, J., Igel, C.: The German traffic sign recognition benchmark: a multi-class classification competition. In: The 2011 International Joint Conference on Neural Networks, San Jose, CA, USA, 2011, pp. 1453–1460. https://doi.org/10.1109/IJCNN.2011.6033395
22. Houben, S., Stallkamp, J., Salmen, J., Schlipsing, M., Igel, C.: Detection of traffic signs in real-world images: the German traffic sign detection benchmark. In: Proceedings of IJCNN, August 2013, pp. 1–8 (2013)
23. Møgelmose, A., Trivedi, M.M., Moeslund, T.B.: Vision-based traffic sign detection and analysis for intelligent driver assistance systems: perspectives and survey. IEEE Trans. Intell. Transp. Syst. **13**(4), 1484–1497 (2012). Dec
24. Houben, S.: A single target voting scheme for traffic sign detection. In: IEEE Intelligent Vehicles Symposium (IV). Baden-Baden, Germany, vol. 2011, pp. 124–129 (2011). https://doi.org/10.1109/IVS.2011.5940429
25. Belaroussi, R., Foucher, P., Tarel, J.-P., Soheilian, B., Charbonnier, P., Paparoditis, N.: Road sign detection in images: a case study. In: Proceedings of IEEE International Conference on Pattern Recognition, pp. 484–488 (2010)
26. Stallkamp, J., Schlipsing, M., Salmen, J., Igel, C.: Man vs. computer: benchmarking machine learning algorithms for traffic sign recognition. Neural Netw. **32**, 323–332 (2012). Aug
27. Iparraguirre, O., Amundarain, A., Brazalez, A., Borro, D.: Sensors on the move: onboard camera-based real-time traffic alerts paving the way for cooperative roads. Sensors **21**, 1254 (2021). https://doi.org/10.3390/s21041254
28. Bouaafia, S., Messaoud, S., Maraoui, A., Ammari, A.C., Khriji, L., Machhout, M.: Deep pre-trained models for computer vision applications: traffic sign recognition. In: 2021 18th International Multi-Conference on Systems, Signals & Devices (SSD), Monastir, Tunisia, pp. 23–28 (2021). https://doi.org/10.1109/SSD52085.2021.9429420
29. Alawaji, K., Hedjar, R.: Comparison study of traffic signs recognition using deep learning architectures. In: 2022 13th International Conference on Information and Communication Systems (ICICS), Irbid, Jordan, pp. 442–447 (2022). https://doi.org/10.1109/ICICS55353.2022.9811216
30. Lu, E.H.-C., Gozdzikiewicz, M., Chang, K.-H., Ciou, J.-M.: A hierarchical approach for traffic sign recognition based on shape detection and image classification. Sensors **22**, 4768 (2022). https://doi.org/10.3390/s22134768
31. Li, E., Mei, L.: Traffic sign recognition and retrieval using limited dataset in the wild. In: IEEE 4th Advanced Information Management, Communicates, Electronic and Automation Control Conference (IMCEC). Chongqing, China, vol. 2021, pp. 1588–1594 (2021). https://doi.org/10.1109/IMCEC51613.2021.9482122
32. Liu, Z., Du, J., Tian, F., Wen, J.: MR-CNN: a multi-scale region-based convolutional neural network for small traffic sign recognition. IEEE Access **7**, 57120–57128 (2019). https://doi.org/10.1109/ACCESS.2019.2913882
33. Zheng, Y., Jiang, W.: Evaluation of vision transformers for traffic sign classification. Wirel. Commun. Mob. Comput. **2022**, Article ID 3041117, 14 p. (2022). https://doi.org/10.1155/2022/3041117
34. Cao, J., Zhang, J., Jin, X.: A traffic-sign detection algorithm based on improved sparse R-CNN. IEEE Access **9**, 122774–122788 (2021). https://doi.org/10.1109/ACCESS.2021.3109606
35. Xu, M., Gu, Y., Poslad, S., Xue, S.: Optimized computation combining classification and detection networks with distillation. In: 2021 International Joint Conference on Neural Networks (IJCNN), Shenzhen, China, pp. 1–8 (2021). https://doi.org/10.1109/IJCNN52387.2021.9534331
36. Wang, Q., Zhang, Q., Liang, X., Wang, Y., Zhou, C., Mikulovich, V.I.: Traffic lights detection and recognition method based on the improved yolov4 algorithm. Sensors **22**, 200 (2022). https://doi.org/10.3390/s22010200
37. Fang, H.-F., Cao, J., Li, Z.-Y.: A small network MicronNet-BF of traffic sign classification. Comput. Intell. Neurosci. **2022** (2022). https://doi.org/10.1155/2022/3995209

38. Kim, J., Kang, J.K., Kim, Y.: A low-cost fully integer-based CNN accelerator on FPGA for real-time traffic sign recognition. IEEE Access **10**, 84626–84634 (2022). https://doi.org/10.1109/ACCESS.2022.3197906
39. Güney, E., Bayilmiş, C., Çakan, B.: An implementation of real-time traffic signs and road objects detection based on mobile GPU platforms. IEEE Access **10**, 86191–86203 (2022). https://doi.org/10.1109/ACCESS.2022.3198954
40. Avramović, A., Sluga, D., Tabernik, D., Skočaj, D., Stojnić, V., Ilc, N.: Neural-network-based traffic sign detection and recognition in high-definition images using region focusing and parallelization. IEEE Access **8**, 189855–189868 (2020)
41. Mannan, A., Javed, K., ur Rehman, A., Babri, H.A., Noon, S.K.: Cognition based recognition of partially occluded traffic signs. Scientia Iranica **29**(4), 1881–1897 (2022). https://doi.org/10.24200/sci.2022.55200.4111
42. Atif, M., Ceccarelli, A., Zoppi, T., Gharib, M., Bondavalli, A.: Robust traffic sign recognition against camera failures. IEEE Open J. Intell. Transp. Syst. **3**, 709–722 (2022). https://doi.org/10.1109/OJITS.2022.3213183
43. Mannan, A., Javed, K., Noon, S.K.: Statistical boosting: a preprocessing technique to enhance performance of machine learning and deep learning models on partially occluded traffic signs. In: IEEE 23rd International Multitopic Conference (INMIC). Bahawalpur, Pakistan, vol. 2020, pp. 1–6 (2020). https://doi.org/10.1109/INMIC50486.2020.9318174
44. John, V., Yoneda, K., Qi, B., Liu, Z., Mita, S.: Traffic light recognition in varying illumination using deep learning and saliency map. In: 17th International IEEE Conference on Intelligent Transportation Systems (ITSC), Qingdao, China, pp. 2286–02291 (2014). https://doi.org/10.1109/ITSC.2014.6958056

On the Hardness of Edge Coloring Problems Associated with Scheduling

Michele Barbato, Diego Delle Donne, and Emiliano Lancini

Abstract In this work, we introduce several Edge Coloring problems related to scheduling and we study their computational complexity. In particular, we prove that the Concurrent Open Shop Coloring is NP-hard. This problem can be summarized as a unitary-time Open Shop Problem with a hard time horizon constraint, in which the goal is to minimize the total processing time of the tasks. We prove that this problem is NP-hard by reducing it to a new variant of the edge coloring problem, named the Mono-Polychromatic Edge Coloring. We study the feasibility and hardness of this problem, both for the general case and when the underlying graph is bipartite. We show that the latter case is equivalent to the Vertex Coloring Problem.

Keywords Edge coloring · Open shop · Scheduling

1 Introduction

The *Open Shop Scheduling* problems form a class of optimization problems in which a set of jobs, each composed by tasks, must be scheduled on a set of machines without observing any prescribed order on the execution of the tasks. Depending on the specific problem, several objectives are considered, such as the minimization of the makespan, of the total tardiness, of the number of late jobs, or other criteria commonly adopted in scheduling. Regardless of the objective function, these problems demand

M. Barbato
Dipartimento di Informatica, Università degli Studi di Milano, Milan, Italy
e-mail: michele.barbato@unimi.it

D. D. Donne
IDS Department, ESSEC Business School, Cergy-Pontoise, France
e-mail: delledonne@essec.edu

E. Lancini (✉)
Université Paris Dauphine—PSL, Paris, France
e-mail: emiliano.lancini@lamsade.dauphine.fr

M. Barbato et al. (eds.), *Beyond Frontiers of Operations Research: Emerging Technologies and Innovative Optimization Paradigms*, AIRO Springer Series 14,
https://doi.org/10.1007/978-3-031-85894-9_2

each task be executed each on a pre-specified machine, and each machine can execute at most one task at once. The *Concurrent Open Shop Scheduling* problems allow different tasks of a job to be executed in parallel. When makespan minimization is considered as the objective function, both these problems are known to be NP-hard, as proved by Williamson [11] for the Open Shop Scheduling, and by Roemer [8] for the Concurrent Open Shop Scheduling. The Open Shop Scheduling and Concurrent Open Shop Scheduling problems are motivated by a wide range of applications, including product design and product assembly [1], and airplane maintenance [3, 10] (we detail possible applications at the end of this section). We refer to the monograph of Kubiak [5] for a more extensive discussion. A natural approach to deal with scheduling problems is to transform them into coloring problems on graphs. In particular, this approach has been used to investigate the interactions between Concurrent Open Shop Scheduling and vertex coloring in Ilani et al. [3]. Another natural equivalence exists between Open Shop Scheduling with Unit-time Tasks and the Edge Coloring Problem on bipartite graphs [5].

In our work, we consider the *Concurrent Open Shop Coloring* (COSC), a variant of the Concurrent Open Shop Scheduling problem where tasks have unitary duration. We are given a time horizon, a set of jobs and a set of processors. The unit-time tasks composing each job have to be processed in any order, each task on a pre-specified processor. Processors may have to interact in the time horizon; this operation costs exactly one unit of time that the processors do not use for executing tasks. Our goal is to minimize the sum of the processing times of each job.

We show that COSC is NP-hard, by a reduction from a novel type of edge-coloring problem on bipartite graphs, for which we first give a proof of its NP-hardness.

Applications

COSC can be used to model several problems arising in real-world applications. The first example we present is a general project management problem. In this context, a staff of senior and junior employees is allocated to various projects over a predefined time period, such as a working week. Each employee can work on a single project in any given time slot, and there may be requirements of one-to-one meetings between junior and senior staff members. In the context of COSC, each employee is represented by a processor and each project by a job. The COSC objective function fosters solutions where employees collaborate simultaneously on the same project, which is beneficial to the overall outcome and efficiency of the project itself. This setting can be adjusted to different fields of application, like the airplane maintenance case, discussed in [2]. We further illustrate this setting in the context of academia in which students and supervisors work in different projects. Figure 1 depicts the scheduling of a workday of 4 time slots, in which one supervisor (p_1) and three students (p_2, p_3, and p_4) have to work on three different projects (j_1, j_2, and j_3). The supervisor has to schedule one meeting with p_2 and one with p_3. The solution (bottom table) is represented as an edge coloring of the right-hand graph, where each color represents a time slot.

Our second example concerns a less intuitive application of COSC, related to privacy-focused learning problems, inspired by the *vertical federated learning*

Instance

Activity	People assigned
Project j_1	p_1, p_2, p_4
Project j_2	p_1, p_3, p_4
Project j_3	p_2, p_3
Meeting	p_1, p_2
Meeting	p_1, p_3

Solution

Person \ Time	9-11	11-13	14-16	16-18
p_1	j_1	j_2	Meeting	Meeting
p_2	j_1		j_3	Meeting
p_3	j_3	j_2	Meeting	
p_4	j_1	j_2		

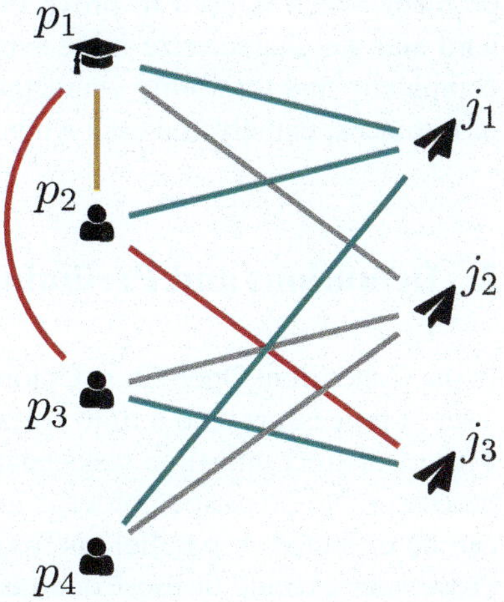

Fig. 1 Scheduling of a workday of 4 time slots in a small team (instance described in the top table). A supervisor (p_1) and three students (p_2, p_3, and p_4) have to work on three different projects (j_1, j_2, and j_3). The supervisor has to schedule one meeting with p_2 and one with p_3. The solution (bottom table) is represented as an edge coloring of the right-hand graph, where each color represents a time slot

framework [6, Chap. 18]. In this context, a *vertical splitting* of a dataset is a partition of the dataset columns (i.e., attributes). Typically, it is performed in such a way that the columns within the same set of the partition do not allow to infer sensitive information (as opposed to using columns in distinct sets of the partition). We consider a set of machine learning algorithms and a dataset. The dataset is split vertically, resulting in a number of smaller datasets associated with column subsets. Each smaller dataset originating from the initial dataset must be assigned to a subset of suitable machine learning algorithms, so to train corresponding predictive models. Given a number k of time slots, a feasible solution consists in establishing those during which each smaller dataset is accessible by the assigned algorithms to read data and perform the training over them. Each algorithm is assumed to be hosted in a computer node, thus it can work with at most one dataset during each time slot. This setting ensures privacy because a machine learning algorithm handles only a subset of attributes which is "privacy-compliant". The objective is to find a feasible solution that minimizes the number of time slots during which each dataset is accessible, so to foster security. This problem can be modeled as a COSC by considering each algorithm as a processor and each (smaller) dataset as a job.

Outline

In Sect. 2 we define the Concurrent Open Shop Coloring both in graph-theoretical terms and as a binary linear program. We discuss basic feasibility results and bounds on the objective function. In Sect. 3 we introduce and study a novel edge-coloring problem which is used to prove that the Concurrent Open Shop Coloring is NP-hard and we characterize its feasibility. To conclude, we study the computational complexity and feasibility characteristics of this problem on a particular subfamily of instances, namely, the case when the underlying graph is bipartite and simple.

2 Definitions and Preliminary Results

In the Concurrent Open Shop Coloring (COSC) problem, we are given a set of jobs, a set of processors, and a time horizon of length $k \in \mathbb{Z}_{>0}$. Each job is composed by a set of distinct unit-time tasks each of which must be executed on a pre-specified processor. No prescribed order is given on the execution of the tasks, and the tasks can be executed in parallel, as long as these are executed in different processors. Processors execute at most one task at a time. Moreover, we allow processors to interact one with another once during the time span. When present, this interaction takes one time unit during which the two processors cannot execute any task. The *processing time* of a job is the number of different (unitary) time slots in which at least one task of the job is being processed. The goal of COSC is to minimize the sum of the processing times of each job.

The following gives a model for COSC stating it as a graph coloring problem. Let V be a set of vertices partitioned into subsets J and P representing the jobs and the machines, respectively. Let $G = (V, E)$ be the graph whose edge-set E is such that:

- for $j \in J$ and $p \in P$, edge jp exists if and only if the job represented by j has a task that must be executed on the processor represented by p;
- for $p, q \in P$, edge pq exists if and only if the two corresponding processors must interact.

Representing each time slot with a color in the set $[k] := \{1, 2 \dots, k\}$, we want to find an assignment of colors to the edges in E such that no two edges incident to a same vertex in P receive the same color. The objective is to minimize the sum over all vertices $j \in J$ of the number of distinct colors used on the edges of $\delta(j) := \{e \in E : e \text{ is incident to } j\}$.

Using the above graph-theoretical representation, COSC can be formalized as a binary linear program as follows. Let x_{ec} be a binary variable equal to 1 if and only if color $c \in [k]$ is assigned to edge $e \in E$, and, for all $j \in J$, let y_{jc} be a binary variable equal to 1 if and only if color c is assigned to at least one edge of $\delta(j)$. With these definitions, COSC amounts to solve:

$$\min \sum_{j \in J} \sum_{c \in [k]} y_{jc} \tag{1}$$

$$\text{s.t.} \sum_{c \in [k]} x_{ec} = 1 \qquad \forall e \in E, \tag{2}$$

$$x_{ec} \leq y_{jc} \qquad \forall e \in \delta(j), \forall j \in J, \forall c \in [k], \tag{3}$$

$$\sum_{e \in \delta(p)} x_{ec} \leq 1 \qquad \forall p \in P, \forall c \in [k], \tag{4}$$

$$y_{jc} \in \{0, 1\} \qquad \forall j \in J, \forall c \in [k], \tag{5}$$

$$x_{ec} \in \{0, 1\} \qquad \forall e \in E, \forall c \in [k]. \tag{6}$$

A straightforward lower bound for COSC is $|J|$. Moreover, the problem is clearly NP-hard, as it contains as a special case the edge-coloring problem (when $J = \emptyset$). This link motivates the following observation.

Proposition 1 *The following hold:*

1. *COSC is always infeasible if $k < \Delta(P)$ and always feasible if $k \geq \Delta(P) + 1$, where $\Delta(P)$ denotes the maximum degree of the vertices in P.*
2. *If every vertex in P is adjacent to at least one vertex in J, then COSC admits a feasible solution if and only if $k \geq \Delta(P)$, and we can build such a solution in polynomial time.*

Proof Clearly, COSC is infeasible if $k < \Delta(P)$. Let G be the graph underlying COSC, and let G' be the graph built by removing each vertex j in J, adding one vertex t for each task t of the job associated to j, and joining this vertex with the vertex associated to the processor that must execute that task. Now, point 1 stems directly from Vizing's theorem [9]: a proper edge coloring on G' corresponds to a feasible solution to COSC.

For point 2, let G'' be the subgraph of G' induced by P. Since every vertex in P is adjacent to at least one vertex outside P, we have that $\Delta(G') \geq \Delta(G'') + 1$. We can properly edge-color G'' with $\Delta(G'') + 1$ colors in polynomial time using Misra and Gries algorithm [7]. Then, we can extend such coloring to G' by greedily coloring the remaining edges. Since each of these edges is incident to a vertex of degree 1, the resulting edge coloring uses exactly $\Delta(G') = \Delta(P)$ colors. □

For the sake of clarity, we have defined COSC assuming that the underlying structure is a simple graph. However, Proposition 1 can be adapted to the case of graphs with parallel edges, relying on Vizing's theorem for such graphs.

Proposition 1 states that a feasible solution to COSC can be found in polynomial time if either $k \geq \Delta(P) + 1$ or if each processor executes at least one task. Consequently, one might initially surmise that COSC can be solved in polynomial time under these circumstances. However, contrary to this expectation, we demonstrate that COSC remains NP-hard, even when these conditions are met. This holds true even when there are no edges connecting vertices in P, in which case the underlying graph is bipartite.

3 The Mono-Polychromatic Edge Coloring Problem

We prove that COSC with $k \geq \Delta(P) + 1$ is NP-hard by considering an auxiliary edge-coloring problem defined as follows.

Let $G = (V, E)$ be a graph, with V partitioned into two subsets of vertices, M and P. The *Mono-Polychromatic Edge Coloring* (MPEC) asks to minimize the number of colors assigned to the edges of G so that for every vertex $v \in P$, all edges incident to v are assigned distinct colors, and for every vertex $w \in M$, all edges incident to w receive the same color. The vertices in P are called *polychromatic*, those in M are called *monochromatic*. To simplify the proofs below we assume that E may contain parallel edges. The main structural difference between the solutions to COSC and those to the MPEC is that in the latter case, all edges incident to a same vertex in M must have the same color. This does not hold for edges incident to a same vertex of J in the former case. Moreover, the graph underlying a MPEC instance may have parallel edges, while that underlying COSC has not.

A graph is said to be *k-MP-colorable* if there exists a solution to MPEC that uses at most k colors. Even for k arbitrarily large, not all graphs admit such a coloration, as it is depicted with the example in Fig. 2: edges w_1v and w_2v must have different colors, however both edges must have the same color as the edge w_1w_2, which is a contradiction. We say that a graph is *MP-colorable* if it is k-MP-colorable for some $k > 0$. Recall that a graph having parallel edges between a monochromatic vertex and a polychromatic one is not MP-colorable.

To characterize MP-colorable graphs we define the following operation: an *M-contraction* consists in contracting an edge e incident to two vertices in M, and assigning to M the resulting vertex. This operation possibly produces parallel edges.

Observation 2 *MP-colorable (resp. non-MP-colorable) graphs remain MP-colorable (resp. non-MP-colorable) after deleting edges between polychromatic vertices and/or M-contracting edges.*

From Observation 2 we reach the following result.

Proposition 3 *A graph G is MP-colorable if and only if it has no circuit containing exactly one polychromatic vertex.*

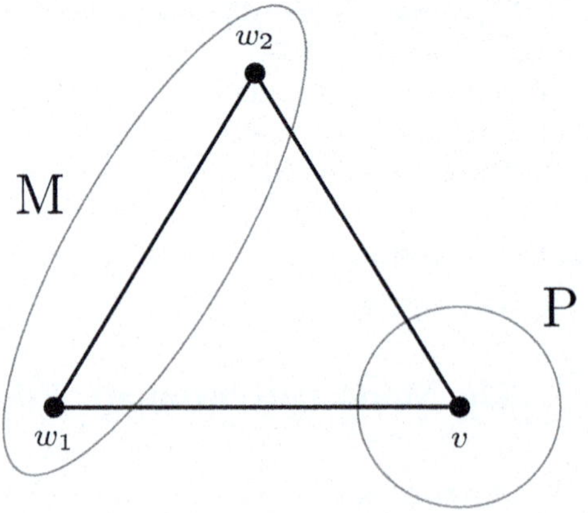

Fig. 2 A graph that does not admit a k-MP-coloring, for any $k \in \mathbb{Z}$

Proof First suppose that G is MP-colorable and has a circuit C composed by only monochromatic vertices except for one polychromatic vertex v. Then, by repeatedly contracting all edges connecting monochromatic vertices in C, we obtain a circuit of length 2 $\{v, w\}$, with $v \in P$ and $w \in M$. By the condition on the polychromatic vertex v, the two parallel edges of the M-contracted circuit must have different colors, but this contradicts the condition on the monochromatic vertex w. Therefore, this contracted instance is not MP-colorable and, by Observation 2, neither is G.

On the other hand, suppose that G has no circuit containing exactly one polychromatic vertex. Let G' be the graph obtained by applying all possible M-contractions and deleting all edges between polychromatic vertices. By Observation 2, if G' is MP-colorable, then so is G. We have that G' is bipartite, as there are no edges between vertices in M (resp. in P).

We have that G is MP-Colorable if and only if G' is simple, as we now show. If G' has two parallel edges, then it is not MP-colorable, hence the same holds for G. If G' is simple, let $v_1, v_2, \ldots, v_m$ be the monochromatic vertices of G'; assigning color i to all edges in $\delta(v_i)$, for all $i = 1, 2, \ldots, m$ yields an MP-coloring that uses m colors.

Therefore to conclude the proof, we suppose by contradiction that there are two parallel edges e and f in G' incident to a vertex $v \in P$. Graph G has no parallel edges between a monochromatic and a polychromatic vertex as otherwise it would be a circuit with exactly one polychromatic vertex. Hence, there are two vertices in M, say u and w, that are endpoints of e and f respectively. Since e and f are parallel, vertices u and w have been identified by the contraction of edges linking monochromatic vertices of G. Such edges and their endpoints form a connected subgraph of G containing u and w. Hence, in this subgraph there is a path $u, v_1, \ldots, v_h, w$ in G such that all v_i are monochromatic vertices. But then, $v, u, v_1, \ldots, v_h, w, v$ is a circuit with exactly one vertex in P, a contradiction. □

As a byproduct of the proof of Proposition 3, we have that bipartite simple graphs with shores M and P where M (resp. P) contains precisely the monochromatic (resp. polychromatic) vertices are always MP-colorable. Nonetheless we now show that finding the minimum k for which such a graph is k-MP-colorable is theoretically difficult. In the following we denote by b-MPEC the latter problem.

In our proof we show that the b-MPEC is equivalent to the *vertex coloring problem* (VCP) on general graphs, that is, the problem of assigning the minimum number of colors to the vertices of a graph so that adjacent vertices are assigned distinct colors.

Proposition 4 *The two following reductions hold true:*

1. *Every instance $\mathcal{V}$ of the VCP admits one instance $\mathcal{M}$ of the b-MPEC such that $\mathcal{M}$ and $\mathcal{V}$ have the same value.*
2. *Every instance $\mathcal{M}$ of the b-MPEC admits one instance $\mathcal{V}$ of the VCP such that $\mathcal{M}$ and $\mathcal{V}$ have the same value.*

Moreover, in each reduction the solution of one problem can be constructed from the solution of the other problem in polynomial time with respect to the size of the instance.

Proof We first prove point 1. Let $G = (V, E)$ be the graph defining instance $\mathcal{V}$ of the VCP. We subdvide every edge $e \in E$ with a new vertex. Let P and F respectively be the sets of new vertices and edges generated by the subdivision. Let us consider the instance $\mathcal{M}$ of the b-MPEC defined over $B = (P \cup V, F)$. An optimal solution to $\mathcal{V}$ gives a feasible solution for $\mathcal{M}$: for every $v \in V$ assign to every edge $f \in \delta(v)$ the color assigned to v in the optimal solution to $\mathcal{V}$. This solution is feasible because a vertex $w \in P$ is incident to precisely the two edges $f_1, f_2 \in F$ obtained by subdividing $e \in E$; since the endpoints of e have distinct colors, also f_1 and f_2 have distinct colors and the condition on vertices in P is satisfied. The condition on the vertices of V (which are monochromatic) is also fulfilled by definition. Conversely, an optimal solution to $\mathcal{M}$ is feasible for $\mathcal{V}$ by assigning to each vertex $v \in V$ the color assigned to its incident edges in that optimal solution. Such a color is unique because the vertices in V are monochromatic, by definition of B, and thus satisfy the condition on monochromatic vertices. Moreover, by the condition on P, adjacent vertices of G receive distinct colors. Therefore point 1 holds.

Point 2 can be proved similarly after replacing with an edge (v_1, v_2) each 2-path (v_1, w, v_2) linking two polychromatic vertices $v_1, v_2 \in P$ of the bipartite graph $B = (P, M; E)$. These operations transform an instance of the b-MPEC into an instance of the VCP (Fig. 3). □

Proposition 4 implies that the Mono-Polychromatic Edge Coloring is NP-hard even when $k \geq \Delta(P) + 1$, or when the underlying graph is bipartite: indeed, we can reduce a vertex coloring problem on a graph with large chromatic number to an instance of b-MPEC where $\Delta(P) = 2$. These facts contrast with classical results on edge coloring problems such as Kőnig's [4], or Vizing's [9]. Finally, we prove that COSC is also NP-hard as it implies the b-MPEC.

Proposition 5 *COSC is NP-hard, even when there exists no edge between processors.*

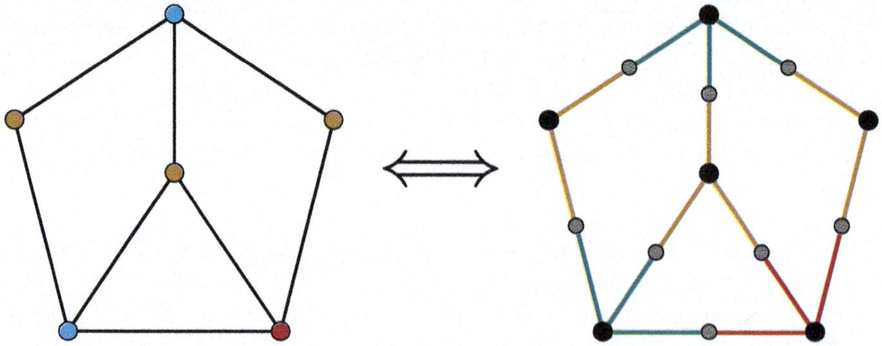

Fig. 3 Equivalence of vertex coloring problem (left-hand graph) and b-MPEC (right-hand graph) via edge subdivision. The gray vertices in the right-hand graph are polychromatic, the black vertices are monochromatic

Proof It is enough to prove the result on the case where there exists no edge between processors. In particular, the underlying graph is bipartite. Let $B = (M, P; E)$ be a bipartite graph, and let $k \geq 0$ be an integer. If B is k-MP-colorable, then the corresponding coloring is also a solution of value exactly $|M|$ to the COSC instance defined on B and k. Since this is the minimum possible value of a solution to COSC, then that solution is optimal.

On the other hand, let x be an optimal solution to COSC on the graph B with at most k colors. If x has value exactly $|M|$, then B is k-MP-colorable, since the edges incident to each vertex $v \in M$ use exactly one color, and hence v is monochromatic. This proves that the problem of deciding whether MPEC admits a solution on $B = (M, P; E)$ of value less or equal than k is equivalent to the problem of deciding whether the COSC defined by $B = (M, P; E)$ and k admits a solution of value exactly $|M|$. Therefore, as MPEC is NP-hard, then also COSC is. □

We remark that the previous proof states that COSC is NP-hard also when there are no interactions between processors, hence we have this result.

Corollary 1 *Minimizing the sum of the processing times of the jobs in the Concurrent Open Shop Problem with tasks of unitary duration is NP-Hard.*

The complexity result of Proposition 5 extends naturally to the cases where the structure underlying COSC is a graph with parallel edges or a hypergraph modelling interactions among multiple processors in the same time slot. Concerning COSC, a research direction is to investigate the analogous of Proposition 1 in the latter case.

Acknowledgements The authors would like to express their gratitude to the anonymous reviewers for their valuable input.

Fundings This work was supported in part by project OECP funded by CNRS—Sciences Informatiques (INS2I).

This work was supported in part by project SERICS (PE00000014) under the NRRP MUR program funded by the EU - NGEU. Views and opinions expressed are however those of the authors only and do not necessarily reflect those of the European Union or the Italian MUR. Neither the European Union nor the Italian MUR can be held responsible for them.

References

1. Ahmadi, R., Bagchi, U., Roemer, T.A.: Coordinated scheduling of customer orders for quick response. Naval Res. Logist. **52**(6), 493–512 (2005)
2. Grinshpoun, T., Ilani, H.: Partially-concurrent open shop scheduling (2014)
3. Ilani, H., Shufan, E., Grinshpoun, T.: Partially concurrent open shop scheduling and graph colourings. In: Proceedings of the 11th international conference of the practice and theory of automated timetabling (PATAT) pp. 195–205 (2016)
4. Kőnig, D.: Über graphen und ihre anwendung auf determinantentheorie und mengenlehre. Mathematische Annalen **77**(4), 453–465 (1916)
5. Kubiak, W.: A Book of Open Shop Scheduling: Algorithms, Complexity and Applications, Volume 325 of International Series in Operations Research & Management Science. Springer International Publishing (2022)

6. Ludwig, H., Baracaldo, N.: Federated Learning: A Comprehensive Overview of Methods and Applications. Springer (2022)
7. Misra, J., Gries, D.: A constructive proof of Vizing's theorem. Inform. Process. Lett. **41**(3), 131–133 (1992)
8. Roemer, T.A.: A note on the complexity of the concurrent open shop problem. J. Sched. **9**(4), 389–396 (2006)
9. Vizing, V.G.: On an estimate of the chromatic class of a p-graph. Diskret. Analiz **3**, 25–30 (1964)
10. Wagneur, E., Sriskandarajah, C.: Open shops with jobs overlap. Eur. J. Oper. Res. **71**(3), 366–378 (1993)
11. Williamson, D.P., Hall, L.A., Hoogeveen, J.A., Hurkens, C.A.J., Lenstra, J.K., Sevast'janov, S.V., Shmoys, D.B.: Short shop schedules. Oper. Res. **45**(2), 288–294 (1997)

Machine Learning Algorithms for Supporting Optimization in Defining Yard Stacking Strategies

Daniela Ambrosino and Haoqi Xie

Abstract In maritime container terminals, the storage yard is recognized as a crucial area in permitting the efficient management of import and export flows. This study focuses on developing the storage rules at a tactical level following the consignment strategy for the export yard based on a European layout. To obtain more rational storage rules, Unsupervised Learning is employed to cluster similar containers. Five algorithms are tested with historical data from an Italian container terminal through Three experimental campaigns. The particular data distribution of containers is analyzed. A comparison of the proposed rule with some benchmark solutions is presented. Some possible tracks to integrate Machine Learning and Operational Research are pointed out.

Keywords Maritime container terminal · Yard storage strategy · Unsupervised learning · Optimization

1 Introduction and Literature Review

Maritime container terminals play a crucial role in the logistic chain, with over 80% of international cargo being transported via sea. Being affected by naval gigantism, the daily throughput of the terminal has been increasing over the years, bringing new issues to enhance the performance of terminal operations, especially efficiency. Therefore, the requirements for novel methods applied to decision-making problems in container terminals emerged.

Among all the areas of container terminals, the storage yard is essential in permitting the efficient management of import and export flows. Furthermore, in recent

D. Ambrosino (✉) · H. Xie (✉)
Department of Economics, CIELI (Centro Italiano di eccellenza sulla logistica, i trasporti e le infrastrutture) University of Genoa, Genova, Italy
e-mail: daniela.ambrosino@unige.it

H. Xie
e-mail: haoqi.xie@edu.unige.it

M. Barbato et al. (eds.), *Beyond Frontiers of Operations Research: Emerging Technologies and Innovative Optimization Paradigms*, AIRO Springer Series 14,
https://doi.org/10.1007/978-3-031-85894-9_3

years, much attention has been devoted to yard management following specific yard template planning [26]. In the storage yard, the main aim is to improve efficiency when performing yard operations, such as storing and retrieving containers and routing/dispatching of material handling equipment, required for transferring containers for their departure to vessels, trains, and trucks.

The storage of containers is widely studied at an operational level, known as the space assignment problem [31]. This problem concerns finding a better location (i.e., a block, a slot, or sub-blocks) for containers abiding by specific storage rules based on some available information. To obtain promising efficiency of yard operations, the main objectives are generally to optimize the containers' handling time [19], the reshuffles [1], the vessels'/trucks' waiting time [29], the equipment traveling distances [11]; in [9], the travel distance is combined with the workload balance of material handling equipment to define the space assignment problem for an import yard, while evaluating the impact on quay cranes performances during the unloading operations, [28] faced the same problem referring to the export yard. According to [4], detailed information on containers, especially the characteristics of containers must be considered in deciding the storage space.

After having defined the template of the yard and the assignment of containers, it is necessary to follow specific stacking strategies to manage the storage of containers in the yard. The storage strategy problem aims to find the best storage strategy in terms of space utilization, equipment deployment, and efficiency in yard operations [4]. The stacking strategies generally differ for the import, export, and rail yards, and for the different types of containers. In the literature, the two strategies commonly adopted are the consignment strategy and the random policy. The random policy, mainly used for import yards and rail yards when information on containers is not available, only satisfies the capacity of the stacks, while assigning containers of the same size (20, 40, or 45 ft) to the available stacks. Adopting this strategy always derives other two operational problems, known as pre-marshaling and re-marshaling. Pre-marshaling aims at reducing reshuffles by reorganizing the stacked containers beforehand while taking into account their handling priority [6]. Re-marshaling attempts to move containers from the current location to the place adjacent to the respective vessel [15]. The idea of the consignment strategy is to store homogeneous containers in the same stacks; this strategy is generally used for export yards. The homogeneity depends on the storage rules that consist of different combinations of the common characteristics of containers, i.e., voyage, size, type, destination, and weight. Such a strategy is proven to require more space than random policy. However, it can help improve the productivity of loading operations for export flows, including pickup, moving, and loading processes performed by material handling equipment [2, 31].

Both simulation techniques and optimization models are used extensively to solve the problems in the storage yard. In recent years, Machine Learning (ML) techniques have also been attracting authors' attention. ML is a branch of artificial intelligence and computer science that uses data and algorithms to imitate how humans learn, gradually improving accuracy (IBM). In [5, 22], the Gaussian Mixture Model (GMM) to cluster containers by their weights is implemented. Reinforcement Learning is employed to determine optimized stacking and retrieving sequence by [7, 12].

[7] apply Q-Learning while [12] use a general reinforcement learning structure. The following works solve hybrid problems: [20] adapt three supervised learning algorithms: multiple linear regression, decision trees, and random forest to predict the dwell time of vessels for supporting the decision of stacking operations; [32] develop a Dynamic Programming model for assigning stacking locations to containers. Results obtained under small-scale instances by a dynamic programming model are treated as input data to a Self-attention neural network (SANN) to predict the solution under large-scale instances. Comparisons with other supervised learning algorithms are also presented.

The main purpose of the present work is to provide new insights into using Machine Learning to support Operational Research in solving the stacking strategy problem in container terminals. Multiple Unsupervised Learning algorithms are tested to cluster containers following the consignment strategy based on historical data. We would like to discuss how Machine Learning can be integrated with optimization models and how to overcome some bottlenecks in formulations. The peculiarity of data distribution and the approaches to managing features are discussed as well.

The remainder of this paper is organized as follows: Sect. 2 will provide a brief problem description; the discussions on using Machine Learning to solve the stacking strategy problem will be put forward in Sect. 3; the experimental results of various campaigns are displayed in Sect. 4; finally, Sect. 5 concludes and outlines future work.

2 Problem Description

The problem under investigation aims to define the best consignment strategy for an export yard characterized by a given layout and reach stackers (RS) and trucks as handling equipment used for transferring containers to the quay-cranes (QC). The layout of the export yard consists of areas, each one with a given number of blocks of different capacities. The capacity of each block is expressed in terms of the number of 20' containers, which corresponds to the number of TEUs, since 1 TEU refers to one 20' container. In this work, the capacity is expressed in terms of the number of containers. The block capacity depends on the number of bays, rows, and tiers composing it [18], as the reader can see in the sketched blocks depicted in Fig. 1; the blue block is composed of 5 bays, 3 rows, and 4 tiers, and the yellow one is composed of 4 bays, 2 rows, and 4 tiers. A set of stacks in a given 20'-bay of a certain block is called a *bay-location*. When RSs are used instead of gantry cranes, it is relevant to refer to the capacity of each bay-location. In the case of the yellow block, each bay-location has a capacity of 8 containers, and a blue block has a capacity of 12 containers. Note that, storing 40' containers requires two contiguous bay-locations.

Let us consider a specific area (composed of blocks of different capacities) dedicated to storing containers waiting for their loading on a specific vessel.

Considering the consignment strategy, general rules set that the containers to be loaded on the same vessel can be stored in the same stack if they have the same port

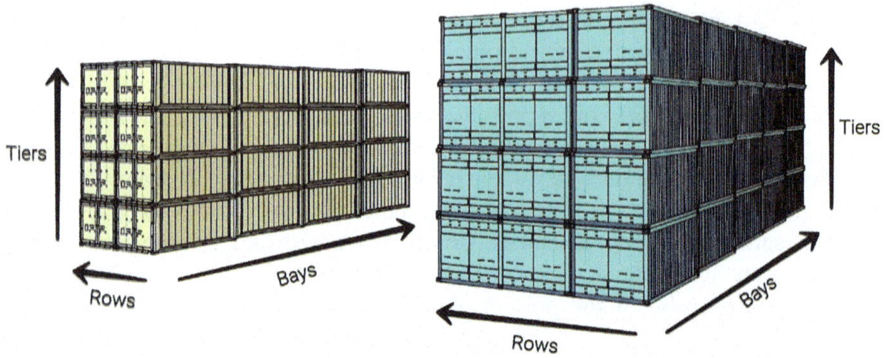

Fig. 1 Two different blocks in an export yard

Table 1 Bay-locations and empty slots in the yard following four storage rules

Information of storage rules and containers per weight class										
	L/H		L/H		L/M/H			L/M/H		
WIs (tons)	0–15	15–33	0–20	20–33	0–15	15–25	25–33	0–17	17–27	27–33
N. Cntrs.	9	11	12	8	4	8	8	9	6	5
Bay-Ls organization and empty slots										
Bay-Ls used	3	3	3	2	1	2	2	3	2	2
Empty slots	4		0		0			8		

Notations 1. L: Light; M: Medium; H: Heavy; WIs: Weight Intervals; Bay-Ls: Bay-locations

of destination, the same size (20', 40', or 45') and type (standard box or High Cube), and similar weights. The principal challenge in finding a better storage rule exists in the definition of homogeneous weights. In the literature, containers are generally grouped into three weight classes: Light, Medium, and Heavy [16]. The number of weight classes will directly influence the number of groups of homogeneous containers to manage, thus, further affecting the number of containers in each group. In addition, another intricate parameter is the weight interval (lower and upper bound) of each weight class (for more details, please refer to [3]). Different combinations of these characteristics will significantly change the bay-locations required to store containers. Let us present a simple case to better clarify this point and understand the impacts on the bay-locations occupancy while adopting 4 different storage rules. Consider having blocks with 20' bay-locations of capacity 4 containers (i.e. one row and 4 tiers) and only 20 containers of 20' to store. Suppose we want to compare the following 4 rules: two weight classes (L/H Light and Heavy) and three classes (L/M/H Light, Medium, and Heavy), both with 2 different weight intervals, as reported in Table 1. Ideally, to obtain the least empty slots, the capacities should be the aliquot parts of the number of containers in each weight class (see the second and the third rules with zero empty slots).

The same Table shows the number of required bay-locations for each group of containers belonging to the different weight classes. The number of bay-locations used ranges from 5 to 7, while the number of empty slots ranges from 0 to 8. Note that the cases with 0 empty slots require both 5 bay-locations, but the strategy that is based on 3 weight classes is preferred from an operational point of view for what concerns the loading operations.

Differently from the simple case presented above, the yard is characterized by a given number of bay-locations of different capacities, that must be assigned to the groups of homogeneous containers obtained from the storage rules.

Finally, note that once the storage rules are defined, they will be adopted in a given time horizon until some modifications are required due to changes in the export flows.

Summarizing, the problem under investigation involves the decision of selecting the promising number of weight classes and their relative weights' upper/lower bounds, and other attributes of containers among multiple alternatives as a storage rule defined with the support of Unsupervised Learning to minimize both the number of bay-locations used and the total empty slots while respecting the yard capacity (i.e., the number of available bay-locations of different capacity).

3 The Discussions on the Use of Machine Learning

When complying with the consignment strategy, it is necessary to focus on two core points: the homogeneity of containers and the management of weights. To handle them properly, appropriate analytic approaches should be adopted. In a recent work: [3] have proposed several methods to solve a similar problem as above-discussed. The definition of rules in [3] is based on historical data analyzed statistically, and the authors divide containers into groups per their destination, types, size, and weights. The proposed model was used to evaluate different rules concerning the weight classes (i.e., the number and the weight limits) derived by the conventional management of the yard.

In the present work, we use Unsupervised Learning (UL) techniques to cluster containers. The algorithms of UL can study the implicit patterns between features (i.e., size, type, destination, and weight) of variables (i.e., containers) in complex data distribution and then aggregate similar data points based on a specific mechanism (please refer to [8]). It can be deemed that the homogeneity of containers and their corresponding weights' lower/upper bounds defined by different UL techniques should be more rational than the common convention since they are obtained based on the historical export flows of the container terminal being studied. UL algorithms can better converge to the real data distribution, especially the number of containers in each weight class, an important element that can influence the space utilization and bay-locations used in the yard as explained in Sect. 2.

The 0–1 programming model proposed by [3] is formulated with a set of complex variables indicating if a bay-location can store containers characterized by a port of destination, a type, size, a class of weight following the weight class configuration chosen. Note that by increasing the attributes used for grouping homogeneous containers, the number of variables in the model will rise as well, which will increase the computational time required to solve the model. Unsupervised Learning might simplify such a decision variable to a modest degree. As a data pre-processing technique, the groups generated by Unsupervised Learning should have already included the above-mentioned attributes. In summary, the new set of variables indicates if a bay-location can store containers belonging to a specific cluster. Such a cluster includes details about size, type, destination, and weight.

Even though UL is strong in performing clustering tasks, its trustworthiness, and correctness should be evaluated. In particular, the impact of weight classes and intervals found by UL algorithms on yard utilization must be observed, amongst which a potential track to integrate Machine Learning and Operational Research: using optimization approaches to evaluate and further improve the Machine Learning-based rules.

As mentioned above, [5], [22] conduct similar investigations but with different purposes and methods. Both works mainly deal with the storage allocation problem at the operational level using online models and approaches. Both works make a pre-processing phase that artificially classifies containers into various groups by their sizes, destinations, types and loading vessels. For each group, the weight of containers among all voyages is calculated. Then, [5], [22] propose to use a probability density-based algorithm, i.e., the GMM [23] to perform the clustering tasks based solely on the weight of containers. To stock containers in the yard, [22] design a Decision Support System embedding the GMM module and an online algorithm module. In contrast, [5] develop an online optimization model and an online algorithm for assigning containers to blocks in real-time.

In the present work, we are interested in a tactical problem. How to define the best rules of the consignment strategy according to different export flows is the principal focus. The defined rules will be proposed and implemented in the terminal Decision Support System used to manage the export containers defining the entering place in the terminal and addressing the trucks to the required bay-location for taking or leaving the container.

3.1 The Selection of ML Algorithms

Five representative algorithms with distinguished mechanisms have been selected and tested with the following motivations:

1. K-means [25]: it is one of the most widely used UL algorithms. Its mechanism is simple in terms of computational complexity. K-means is suitable for most of the clustering tasks with the disadvantage of its sensibility on the noises. Considering that the weight of containers is compacted well between 2.3 and 33 tons, the noises are assumed to be insignificant. Therefore, K-means might perform properly;

2. Spectral Clustering (SP) [27]: there are no special requirements on the types of data (i.e., convex or non-convex) for using SP. It can converge to the global optimum especially if the number of clusters is presumed small. It is granted a huge advantage that this algorithm is based on the eigendecomposition to the similarity matrix between datasets. This mechanism differs from other Euclidean distance-based algorithms, that can deal with sparse data.
3. DBSCAN [24]: differing from K-means and SP, it is not necessary to define the number of clusters a priori. DBSCAN can deal with any shape of data, particularly, dense data to which the weight of containers is similar.
4. Hierarchical Clustering (HC): Agglomerative Clustering [21]: this algorithm can tackle various types of data, especially when the data is hybrid with numerical, categorical, and non-numerical one. Since the characteristics of containers contain different data types, choosing HC is reasonable. Furthermore, its mechanism can be expressed by the dendrogram which is similar to the approach proposed by [3]. Hypothetically, this algorithm might be the most suitable one.
5. Gaussian Mixture Model (GMM): this algorithm is used by two benchmark works: [5, 22]. It is selected to test its performance for the current problem.

To test the five techniques listed above, two computational campaigns have been realized. The first campaign contains a comparison between GMM and other algorithms considering only the weight of containers. The second one considers more attributes of containers (i.e., size, weight, type, and destination) using GMM and the other four algorithms. More details will be presented in the next section.

4 Experiments

The input data is provided by an Italian container terminal. Please note that these data are historical. In total, 9431 containers are involved among which 3679 are 20 ft, 5729 are 40 ft, and 25 containers are 45 ft. All algorithms were coded in Python 3.12 using the *Scikit-learn* library and the experiments were executed on a computer equipped with an Intel Core i7 2.6 GHz CPU and 32 GiB of memory.

4.1 The First Campaign for Clustering

In this campaign, the clustering task is simplified with only one feature, i.e., the weight. It is worth emphasizing that grouping instances by one feature is risky. In the case under investigation, the weight interval is the point of interest and the performance of different algorithms on this point is observed.

Figure 2a shows the distribution of containers considering only weight while Fig. 2b depicts the density diagram. Holistically, the distribution of data is relatively uniform. This distribution does not form visible clusters. Some patterns can

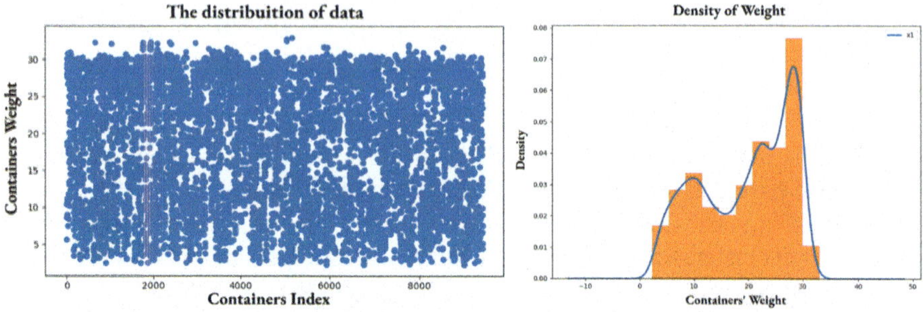

(a) The data distribution. (b) The density diagram.

Fig. 2 The data distribution and density

be noted that the data in the interval around 15 tons is sparse while they are densely agglomerated in the range from 25 to 30 tons. The two-hump density curve validates the similar distribution.

For many UL algorithms, the number of clusters, denoted as *K* value is a hyperparameter that must be decided properly. Generally, the determination of this value relies on different metrics. Examples are illustrated in Fig. 3, where the proposed K is evaluated. Calculating the Sum of Squared Error (the Elbow method) and the Silhouette score can meet the requests of most cases. The principal metrics for GMM are the BIC (Bayesian Information Criterion) and the AIC (Akaike Information Criterion) [17]. The ideal K value can be identified by analyzing the dendrogram shown in Fig. 3b by the HC algorithm that does not require the K hyperparameter. The dendrogram only shows how the data points are agglomerated and implies which clusters are most similar.

Please take a look at Fig. 3a, which depicts the number of components (i.e., K value) and their corresponding BIC scores. Note that the BIC score decreases significantly with the increase of components from 2 to 3, and then remains stable. This means that the goodness-of-fit of the GMM model under the involved dataset approaches optimal when the K is chosen 3. We adopt the Elbow method and the Silhouette score for the K-means method, which obtain distinct results. As presented in Fig. 3c and d, the knee point suggests 5 as the ideal K value while the maximum Silhouette score is dropping on K equaling 3. Please note that we only consider one feature in this campaign. Due to the different mechanisms (please refer to [30] for the Elbow method and [13] for the Silhouette score), the recommended K for K-means might be various, but can be set between 3 and 5. Therefore, we finally select the medium value as 4. Figure 3b illustrates the clusters generated by HC. Each branch represents a new cluster. The higher the branch is, the more notable diversity the cluster has. Hence, selecting one, two, or four as K value for HC is possible. According to the requirement of the problem being studied, 4 is more suitable. At last, K equaling 3 and 4 are tested for the SP method since it is advantageous in fewer clusters.

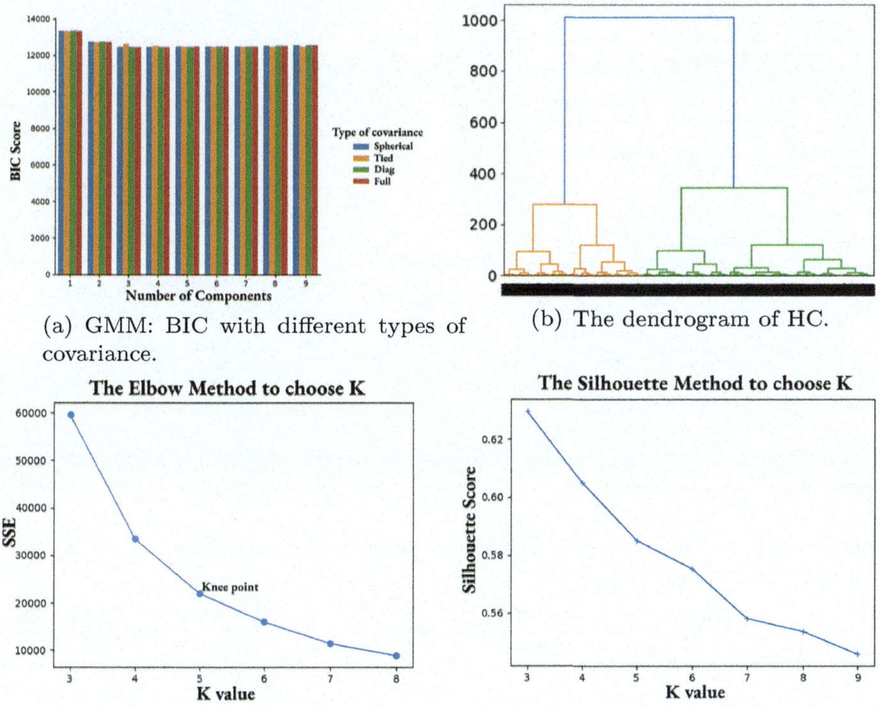

(a) GMM: BIC with different types of covariance.

(b) The dendrogram of HC.

(c) K-means: The Elbow method.

(d) K-means: The Silhouette Score.

Fig. 3 The four approaches for defining K value in different algorithms: the first campaign

DBSCAN is highly affected by two hyperparameters: the searching radius (*eps* and the minimum data points in this range. Figure 4 depicts the relation among these two hyperparameters and K. Using a general *eps* value from 0.2 to 0.6 (see Fig. 4a), only three combinations of hyperparameters can find two clusters. To adjust the eps value, we reconsider the distribution of data. Since the data points (i.e., the weights of containers) are normally distributed between 2.3 tons to 33 tons, the values of adjacent points are extremely similar, thus smaller eps values perform better. Figure 4b presents the clusters obtained when the *eps* value is chosen between 0.025 and 0.2: the distinctiveness dramatically increases. DBSCAN can generate one cluster or almost 300 clusters using pre-defined smaller eps values. In conclusion, for this densely compacted and featureless data, it is challenging for DBSCAN to satisfy the request of the ongoing campaign.

In summary, Table 2 shows the different clusters and their weight intervals generated by various algorithms. DBSCAN is excluded as a consequence of its inappropriate mechanism. It can be noted that GMM and SP define different weights' lower and upper bounds. The upper bounds found by SP are more stressed. The remaining algorithms generate four clusters with very similar bounds. The impact of different

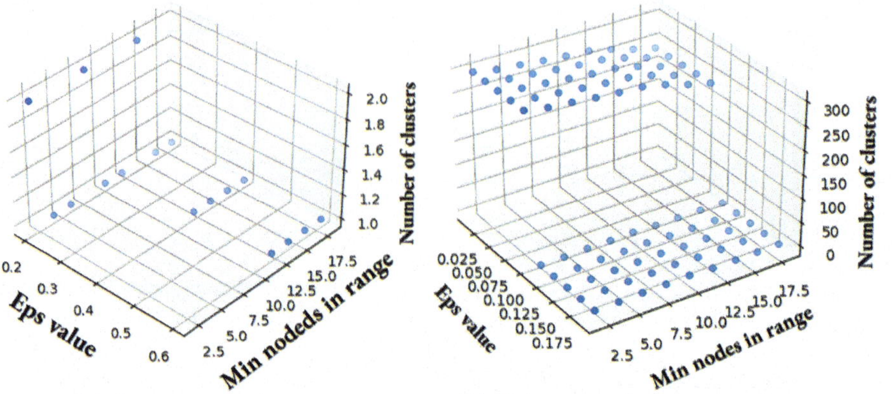

(a) General eps value. (b) Extremely minuscule eps value.

Fig. 4 The change of eps, minimum nodes, and the number of clusters: data and one feature considered in the first campaign

Table 2 The comparison of weight intervals obtained by different algorithms

Groups	Weight interval (tons)				
	GMM	K-means	HC	SP-3 groups	SP-4 groups
Group 1	2.3–15.6	2.3–10	2.3–9.6	2.3–14	2.3–8.4
Group 2	15.7–25.8	10.1–17.2	9.7–17.7	14.1–24.4	8.5–16.7
Group 3	25.9–33	17.3–24.6	17.8–24	24.5–33	16.8–24.6
Group 4		24.7–33	24.1–33		24.7–33

algorithms on the results of this simple task is very subtle. Therefore, it requires more experiments to evaluate which algorithm will outperform.

4.2 The Second Campaign for Clustering

It can be seen from the previous section that one feature can not explain the distribution of data well. In this experimental campaign, the above-viewed algorithms are tested with four standard container attributes as features, i.e., size, type, weight, and destinations.

The destination feature is reorganized following the One-Hot Encoding. One-hot encoding is a general and step-wise technique before using complicated methods to deal with categorical features. It generates a vector where the value of one component is one and of others is zero [10]. Please note that for size, and type features, even though they are numerical features, they still have non-numerical functionality indicating if a container is a normal box, a 40' container, or a High Cube. However,

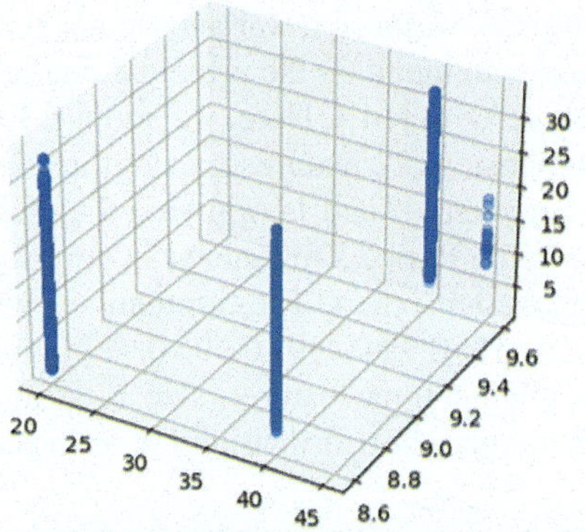

(a) Data distribution based on size, type, and weight.

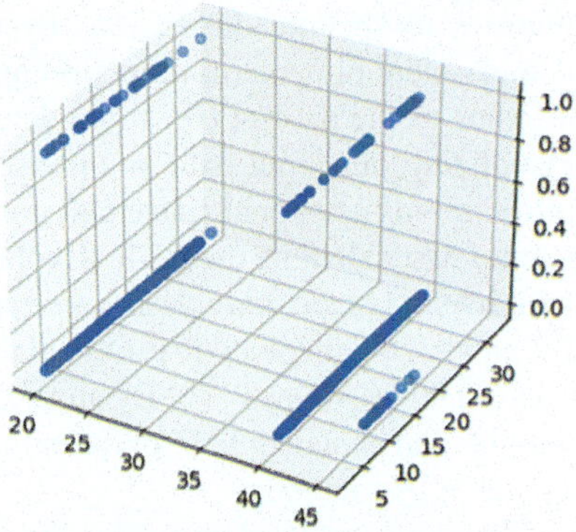

(b) Data distribution based on size, weight, and one destination.

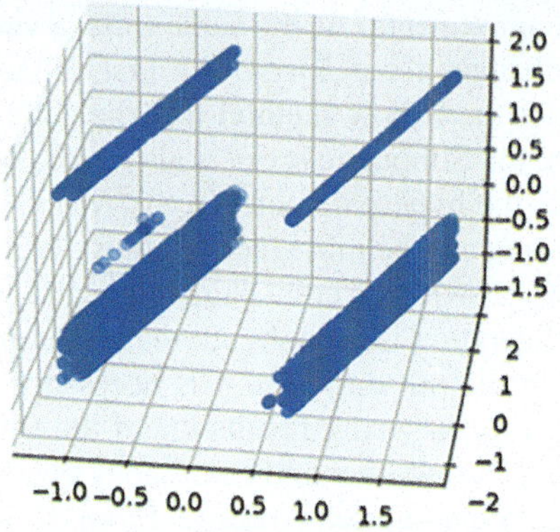

(c) The distribution of data handled by the PCA.

Fig. 5 The data distribution with four container attributes

we maintain the numerical form for these features. The input data is finally prepared with 22 features.

Analyzing the data structure, its particularity can be noted: One-hot encoding has created a large number of one-zero matrix, which occupies the majority of features.

On the other hand, the features of size and type further discretize the data in the space. The only continuous feature is the weight, which significantly impacts data distribution. Compared to the first campaign, the introduction of more features changes the alignment of data in the space as Fig. 5a and b shows.

The form of data is no longer compactly packed. Still, it can be included as the following definition: multiple columnar sets of continuous data are distributed discretely in the different dimensions of space. This structured type of data is caused by fewer continuous features and huge amounts of sparse 0–1 features.

To further analyze the input data, the Principal Component Analysis (PCA) method has been implemented. The PCA method can project high-dimensional features to specific dimensions by calculating the variance-covariance matrix. It is a classical dimension-reduction approach and can maintain the principal characteristics of the dataset (for more details, please refer to [14]). Here, we adopt it for visualizing our input data in an observable dimension. 22 features are compressed into three-dimensional space and normalized between −1 and 1. Applying the PCA should first calculate the EVR (Explained Variability Ratio). This value can describe which features are more descriptive and significant than others. The cumulative EVR for most of the involved features are as follows: (0.616 0.994 0.995 0.996 0.997 0.997 0.998 0.998 0.999... 1. 1.). By analyzing them, the first three features (size, type, and weight) can explain 99.5% of the data. Figure 5c depicts the data distribution reshaped by the PCA method.

The same approaches used in the first campaign are adopted to define the K value. The results are displayed in Fig. 6. It is worth seeing that as the number of components increases, the line trend in Fig. 6a tends to stabilize. Two clear troughs are situated at 6 components and 13 components. At the same time, a vast gap exists when the number of components changes from 2 to 4. We first define 4 as the K value for GMM. For K-means, both the Elbow method and the Silhouette score suggest 4 clusters. In this campaign, the Silhouette score is also tested for SP. Figure 6e shows that 5 clusters are ideal as the hyperparameter in SP. Analyzing the dendrogram, it can be implied that 5 clusters are the most representative of the HC algorithm.

We finally tested DBSCAN with different pairs of *eps* and minimum data points in the radius. A further consideration is brought out in this campaign. Since DBSCAN will detect noise nodes, in the present case, it is necessary to cluster all containers. Therefore, observing the clustering results under the status with zero noise is essential. The results display a smoother and continuous trend in Fig. 7a while b concentrates on the pairs in which no noise is tracked. From (b) we can notice only one pair that generates 5 clusters with no noise. The major pairs classify containers into 3 groups. Hence, we tested these two classification rules whose results are illustrated in Fig. 8. This algorithm is still not able to classify the data effectively. With five clusters, it only defined the outliers concerning their related sizes as other groups. This result supported the previous observation, that DBSCAN can not handle the current input data in a flat and dense arrangement.

Likewise, detailed information on clusters obtained by different algorithms is presented in Table 3. With the introduction of more features, the four algorithms manifest distinct performances. K-means and HC are based on the Euclidean distance between nodes and centroids while SP relies on graphic theory and expresses the relationship between data points by the similarity matrix. The weights can have a great impact on the computation of HC and K-means. Overlapping of weights among clusters exists in HC, SP, and GMM. K-means can exclusively distinguish the thresholds of different clusters with sharp detection. SP can distinguish two similar sizes: 40' and 45', even though there are only 25 units of 45' containers in the data set. The weight intervals for classifying 20' containers found by SP differ from other algorithms.

An interesting anomaly can be outlined in the results of GMM. This algorithm is incapable of clustering the weights of containers. GMM simply groups containers

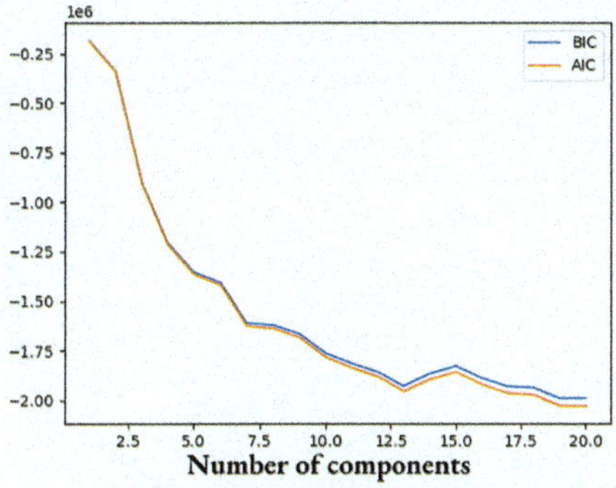

(a) GMM: BIC and AIC score.

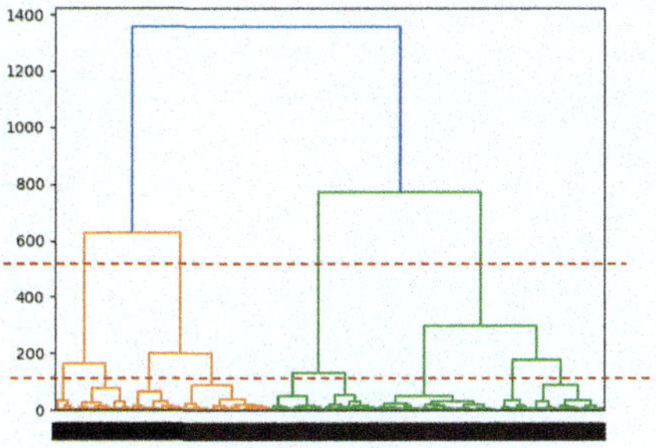

(b) The dendrogram of HC.

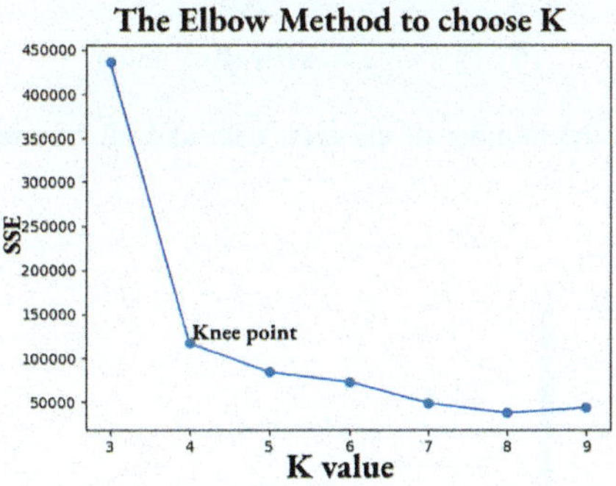

(c) K-means: The Elbow method.

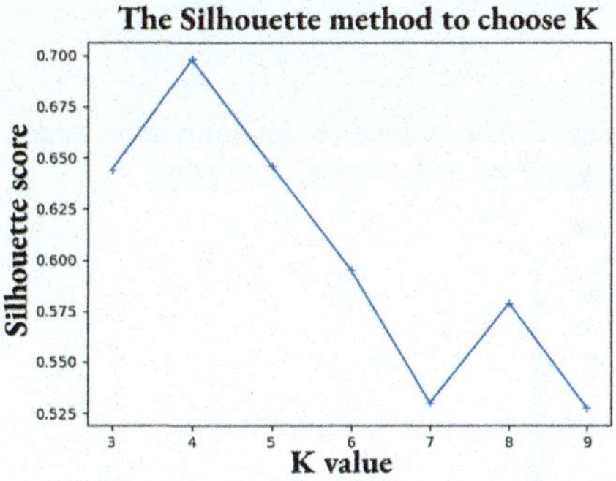

(d) K-means: The Silhouette Score.

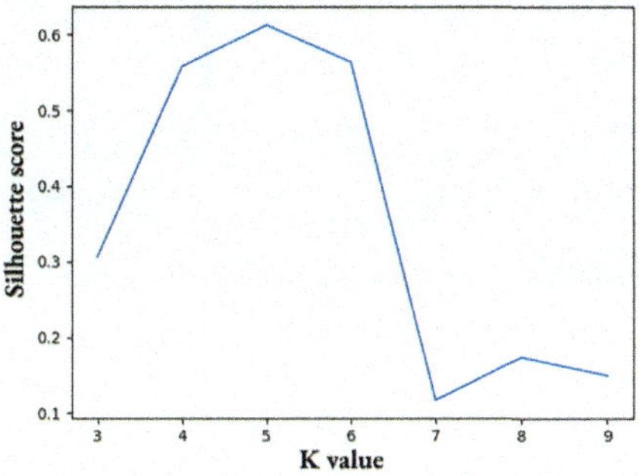

(e) SP: The Silhouette Score.

Fig. 6 The four approaches to define K value in different algorithms: the second campaign

based on the 20 and 40 ft (i.e., the feature of size). To further understand the reason behind this behavior, we check the features related to the destinations which are hardly explicable. Focus has lied on two groups of 20 ft (i.e., 2.6–33, 20 ft and 2.3–30.5, 20 ft), denoted as Group 3 and Group 4 to observe if their destinations have disrupted the clustering results. Results are presented in Fig. 9 where 0 means that in the considered cluster, containers do not have this destination (i.e., D1, D2,..., D19) while 1 is the opposite situation. This figure illustrates that the destinations

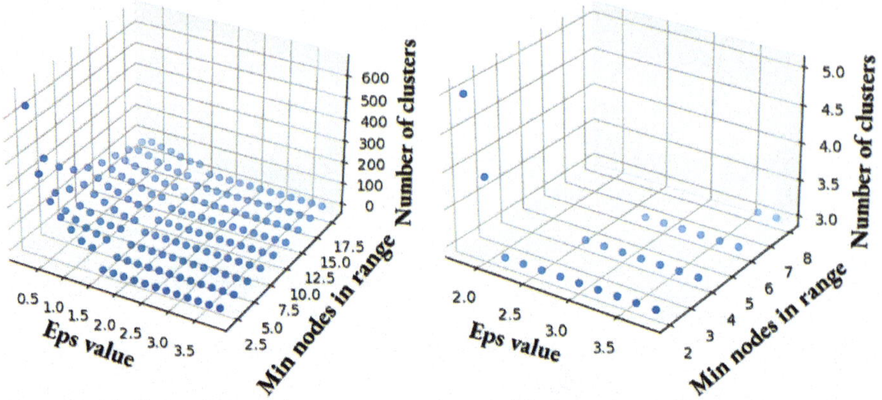

(a) General Eps values. (b) Eps values with zero noise.

Fig. 7 The change of eps, minimum nodes, and the number of clusters: data and all features considered in the second campaign

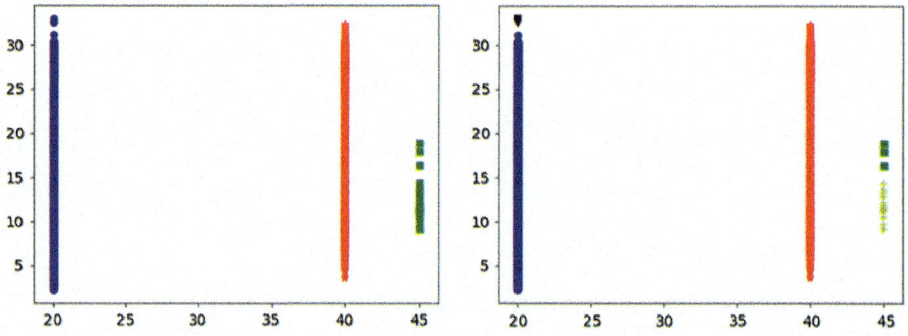

(a) DBSCAN: 3 clusters (b) DBSCAN: 5 clusters

Fig. 8 The results of DBSCAN based on two numbers of clusters

in these groups are complementary. This is because GMM exclusively categorizes 20 ft containers into two groups and applies the same approach to 40 ft containers. This simple categorization is disrupted by the primary characteristics of containers: their length (20 or 40 ft) and height (8.6 or 9.6). GMM assumes a new container belongs to one of these four categories to maximize the probability of the cluster assignment. In this case, the features of destinations are hardly influential. However, if the components (K value) increase, the clusters will be segmented further based on the destinations because this set of features can meticulously categorize the containers even more, and gradually converge to fit the data distribution shown in Fig. 5.

Table 3 Clusters generated by different algorithms

Groups	K-means			HC			SP			GMM		
	Weight	Sizes (ft)	Types	Weight	Sizes (ft)	Types	Weight	Sizes (ft)	Types	Weight	Sizes (ft)	Types
Group 1	3.7–19.3	40/45	8.6/9.6	3.7–15.9	40/45	8.6/9.6	3.7–19.2	40	8.6/9.6	4.5–32.4	40/45	8.6/9.6
Group 2	19.3–32.4	40	8.6/9.6	14.4–27.6	40/45	8.6/9.6	19.1–32.4	40	8.6/9.6	3.7–32.4	40/45	8.6/9.6
Group 3	2.3–15.3	20	8.6	26–32.4	40	8.6/9.6	9.2–18.5	45	9.6	2.6–33	20	8.6
Group 4	15.4–33	20	8.6	2.3–16.7	20	8.6	2.3–14.2	20	8.6	2.3–30.5	20	8.6
Group 5				15.9–33	20	8.6	14.1–33	20	8.6			

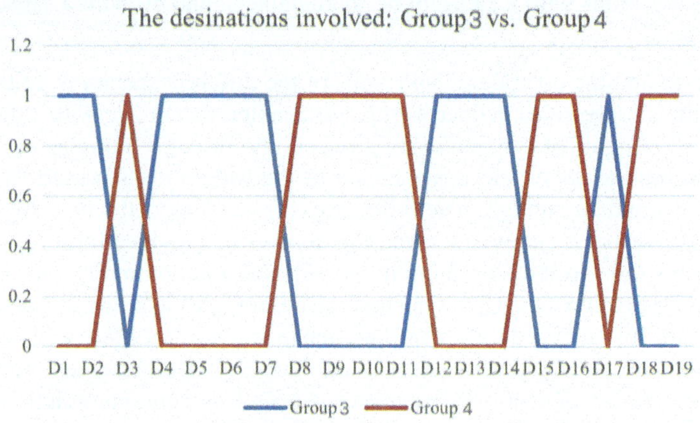

Fig. 9 The destinations considered in Groups 3 and 4

4.3 The Third Campaign: Integrating ML and OR

The previous two campaigns have shown the clustering results. In this campaign, we integrate the operational research technique with UL to evaluate how these storage rules (i.e., different clusters and their corresponding weights' upper/lower bounds) influence the space utilization in the container yard. In particular, the model proposed in [3] is chosen for the experimental tests. Please note that this model uses variables to manage the allocation of containers characterized by a specific size, type, and destination to a bay-location. Thus, the idea was to pass to the model the clusters obtained as results of the first campaign. The model is implemented in Python 3.12 on the same device for testing UL algorithms. Gurobi 11.0 is used as a commercial solver.

Two different scales of instances are generated. The small-scale instance is chosen with 86 containers with the same destination, while 321 containers with three destinations are selected as a medium-scale instance. For both instances, we have considered that the yard has enough space for storing containers. Furthermore, to comprehend how the yard configuration (i.e., the capacities of bay-locations and their corresponding availability) impacts the results, for the small-scale instance, we implement two different configurations. The details of the experimental tests are shown as follows:

1. Scenario 1: the small-scale instance is tested. 24 bay-locations can be used, 8 bay-locations for each of the following capacities: 12, 16, or 20 containers.
2. Scenario 1_bis: the small-scale instance is tested. 20 bay-locations can be used, 10 bay-locations for each of the following capacities: 16, and 20 containers.
3. Scenario 2: the medium-scale instance is tested. 80 bay-locations can be used, 40 bay-locations with a capacity of 16 containers, and 40 with a capacity of 20 containers.

Table 4 shows the computational results of the different scenarios. In the second column, the storage rules defined by [3] are denoted as *DA Xie* with suffixes indicating if the weights of containers are grouped into three or four classes. The third column shows the lower and upper bound of the weight classes used when defining the consignment strategy. The following columns report the objective function value, the number of empty slots, and the number of bay-locations used. The last two columns refer to the CPU time in seconds and the optimality gap. Figure 10a illustrates the details of the CPU time using different rules in scenario 1 and scenario 1_bis. Figure 10b depicts the same information in scenario 2. Likewise, Fig. 11 shows the empty slots caused and bay-locations used under different scenarios respectively.

It can be noted that the solved model can find the optimal solutions in both scenarios. All instances of scenarios 1 and 1_bis (that differ for the weight classes storage rules) have been solved up to optimality within less than 10 s. The CPU times in scenario 2 (i.e., 321 containers) are higher, especially for the storage rules obtained by UL algorithms. The CPU times in scenario 1 are generally superior to those in scenario 1_bis due to the involvement of three different capacities (12,16, and 20).

As for the empty slots, the overall trends in scenario 1, 1_bis, and scenario 2 remain similar. The empty slots caused based on four categories of weight classes (i.e., K-means, HC, SP-4, DA Xie 4a, and Da Xie 4b) exceed those based on three categories notably. For all scenarios, the minimum empty slots are found by adopting the storage rule generated by SP, which are 40, 70, and 191. For scenarios 1 and 1_bis, the bay-locations used are almost the same except for the rules by adopting SP. The optimal number of the bay-locations obtained in scenario 2 is 31, which is found by storage rules Da Xie 3b and SP. One anomaly appears on GMM where 41 bay-locations are used, which even surpasses other results.

Confronting scenario 1 and 1_bis, scenario 1 found an optimal solution characterized by 40 empty slots thanks to bay-locations of 12 containers. Nevertheless, its corresponding bay-locations used are the highest equaling 15. In such a situation, bay-locations within 12 capacities are used more, and the number of containers in

Table 4 The computational results of two scenarios under different storage rules

Scenarios	Index	Weights' bounds (tons)	Objective	Empty slots	Bay-locations	CPU time(s)	GAP(%)
Scenario 1	DA Xie 3a	0–12/12–22/22–33	55	46	9	2.48	0
	DA Xie 3b	0–15/15–25/25–33	55	46	9	3.48	0
	GMM	2.3–15.6/15.7–25.8/25.9–33	60	50	10	2.83	0
	SP-3	2.3–14/14.1–24.4/24.5–33	55	40	15	3.57	0
	K-means	2.3–10/10.1–17.2/17.3–24.6/24.7–33	106	94	12	2.72	0
	HC	2.3–9.6/9.7–17.7/17.8–24/24.1–33	110	98	12	2.57	0
	SP-4	2.3–8.4/8.5–16.7/16.8–24.6/24.7–33	106	94	12	3.19	0
	DA Xie 4a	0–12/12–20/20–28/28–33	85	74	11	5.12	0
	DA Xie 4b	0–10/10–18/18–25/25–33	106	94	12	5.23	0
Scenario 1_bis	DA Xie 3a	0–12/12–22/22–33	79	70	9	2.86	0
	DA Xie 3b	0–15/15–25/25–33	79	70	9	3.31	0
	GMM	2.3–15.6/15.7–25.8/25.9–33	96	86	10	2.28	0
	SP-3	2.3–14/14.1–24.4/24.5–33	79	70	9	2.51	0
	K-means	2.3–10/10.1–17.2/17.3–24.6/24.7–33	138	126	12	1.44	0
	HC	2.3–9.6/9.7–17.7/17.8–24/24.1–33	142	130	12	1.43	0

(continued)

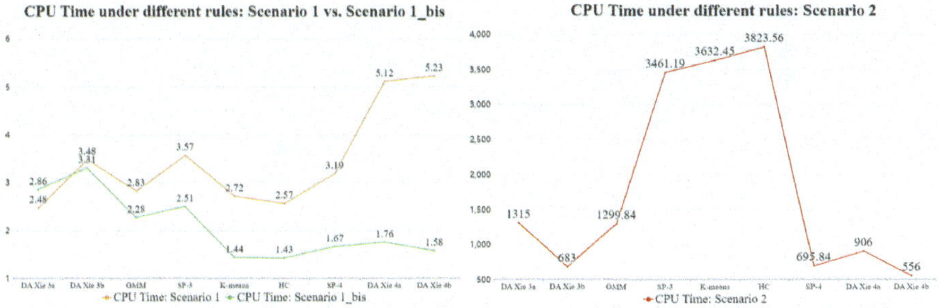

(a) The CPU time under different rules: scenario 1 and scenario 1_bis.

(b) The CPU time under different rules: scenario 2.

Fig. 10 The results of CPU time in different scenarios

each weight's classes defined by the SP-3 rule may be or approximate to a multiple of twelve respecting other capacities (16 or 20). This phenomenon points out that the yard configuration is crucial. If the defined storage rules whose data distribution in each weight's class can fit the yard configuration (particularly, the capacity) well, the empty slots and the bay-locations can be managed better.

Table 4 (continued)

Scenarios	Index	Weights' bounds (tons)	Objective	Empty slots	Bay-locations	CPU time(s)	GAP(%)
	SP-4	2.3–8.4/8.5–16.7/16.8–24.6/24.7–33	138	126	12	1.67	0
	DA Xie 4a	0–12/12–20/20–28/28–33	117	106	11	1.76	0
	DA Xie 4b	0–10/10–18/18–25/25–33	138	126	12	1.58	0
Scenario 2	DA Xie 3a	0–12/12–22/22–33	255	223	32	1315	0
	DA Xie 3b	0–15/15–25/25–33	238	207	31	683	0
	GMM	2.3–15.6/15.7–25.8/25.9–33	235	194	41	1299.84	0
	SP-3	2.3–14/14.1–24.4/24.5–33	222	191	31	3461.19	0
	K-means	2.3–10/10.1–17.2/17.3–24.6/24.7–33	378	339	39	3632.45	0
	HC	2.3–9.6/9.7–17.7/17.8–24/24.1–33	378	339	39	3823.56	0
	SP-4	2.3–8.4/8.5–16.7/16.8–24.6/24.7–33	399	359	40	695.84	0
	DA Xie 4a	0–12/12–20/20–28/28–33	356	319	37	906	0
	DA Xie 4b	0–10/10–18/18–25/25–33	377	339	38	556	0

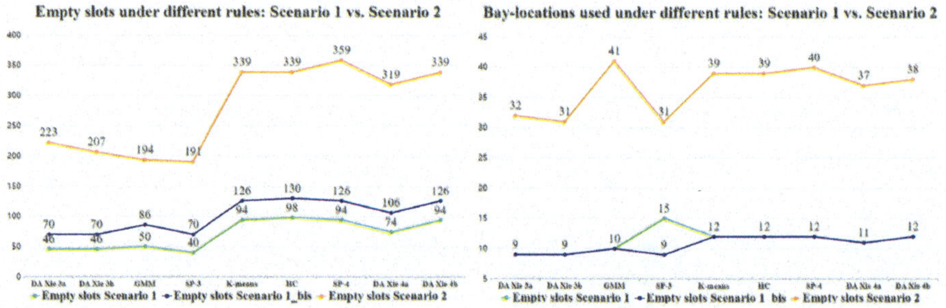

(a) The empty slots caused by using different rules.

(b) The bay-locations used by adopting different rules.

Fig. 11 The empty slots and bay-locations obtained in different scenarios

The above-discussed reason can also explain the anomaly of the GMM rule that appears in scenario 2, where the most bay-locations are used, but with a suboptimal value of empty slots equaling 194. Let us first compare two benchmark solutions where the empty slots are reduced from 223 to 207 with a difference of 16. Their bay-locations used are reduced by one. This implies that one bay-location with a capacity of 16 is not in utilization following DA Xie 3b. Secondly, we focus on the GMM rule and DA Xie 3b whose weight bounds are very similar. However, the results are extremely distinguished. As explained in Sect. 2, different storage rules will cause significantly variegated container distribution in each weight class. The number of containers that may have originally converged on multiples of 20 adopting DA Xie 3b will now follow a pattern that may approximate multiples of 16 abiding by the GMM rule (the same way for the switch from 16 to 12), which causes more bay-locations and fewer empty slots. This representative result between two similar rules has indicated that more detailed weight intervals considering the real export flows shall perform better than general ones.

It is worth noting that the model used in this Section contains a set of variables that determine the allocation of containers with a specific combination of attributes. The exceptionally high CPU times (SP-3, K-means, and HC) could be imputed to the reason that based on these rules, the combination of attributes in each weight class varies greatly, leading to a significant increase in the computational complexity.

In conclusion, the three-categories storage rules are overall promising, especially the one defined by the SP algorithm. It outperforms the benchmark rules proposed in [3] in terms of empty slots. One reason for which this method stood out is that it is minimally affected by the distances between data points. Instead, it can learn better the pattern among features and datasets relying on the similarity matrix. In the scenarios described in this problem, the volume of data and the number of clusters required are both relatively small, which is well suited to the mechanism of the SP algorithm. Another benchmark algorithm: GMM performs suboptimally because its mechanism is based on the probability density, which is indifferent to the distance between data points as well. However, we have to restate that undertaking the clustering tasks using one feature is dull and risky. It is urgent to establish a new mathematical programming model to further evaluate the results presented in the second campaign. Furthermore, it needs to be clarified that the input data tested in this campaign are partially selected from the complete data used for training UL algorithms. The classification results presented in the first and the second campaigns do not perfectly describe the data distribution of 86 and 321 containers, but they can serve as a partial reference. Even so, the model still found new optimal solutions confronting results in the literature, which can prove that ML techniques (UL algorithms) have the potential to support and improve problem-solving conducted by operations research techniques.

5 Conclusion and Future Work

This work attempts to deal with a tactical problem in the storage yard of a container terminal by integrating Machine Learning and Operational Research. The present paper focuses on the Machine Learning phase to cluster homogeneous containers and determine better weight classes and weights' lower/upper bounds. Five algorithms are tested considering one feature or more features respectively. The clusters obtained by Machine Learning algorithms are then tested in a MILP model from the literature. The preliminary results seem to be reasonable. The particular data distribution constructed by the attributes of containers is also discussed. The adaptability of different algorithms to this data distribution is shown as well. By using Machine

Learning-based clusters as input data for the mathematical programming model, the present work has provided a possible track to integrate these two techniques.

In future work, from the perspective of UL algorithms, special observation should be set on the Spectral Clustering since it is outstanding according to the current results. Furthermore, how to manage the features of destinations in an appropriate way for UL algorithms will be another future research focus. As far as the optimization is concerned, it is necessary to test the storage rules acquired by UL algorithms considering all features and evaluate their effectiveness on space utilization in the container yard. For this scope, a new mathematical model is under investigation. On a more practical level, for yard management, the optimal yard configuration and the number of bay-locations should be studied as well.

References

1. Ahmt, J., Sigtenbjerggaard, J.S., Lusby, R.M., Larsen, J., Ryan, D.: A new approach to the container positioning problem. Flex. Serv. Manuf. J. **28**, 617–643 (2016)
2. Ambrosino, D., Sciomachen, A.: Impact of yard organisation on the master bay planning problem. Marit. Econ. Logist. **5**(3), 285–300 (2003)
3. Ambrosino, D., Xie, H.: Optimization approaches for defining storage strategies in maritime container terminals. Soft Comput. **27**(7), 4125–4137 (2023)
4. Carlo, H.J., Vis, I.F., Roodbergen, K.J.: Storage yard operations in container terminals: literature overview, trends, and research directions. Eur. J. Oper. Res. **235**(2), 412–430 (2014)
5. Cho, S.W., Park, H.J., Kim, A., Park, J.H.: Gmm-based online optimization for container stacking in port container terminals. Comput. Ind. Eng. **173**, 108671 (2022)
6. Expósito-Izquierdo, C., Melián-Batista, B., Moreno-Vega, M.: Pre-marshalling problem: heuristic solution method and instances generator. Expert Syst. Appl. **39**(9), 8337–8349 (2012)
7. Fechter, J., Beham, A., Wagner, S., Affenzeller, M.: Approximate q-learning for stacking problems with continuous production and retrieval. Appl. Artif. Intell. **33**, 1—19 (2018)
8. Ghahramani, Z.: Unsupervised., learning. Advanced Lectures on Machine Learning: ML Summer Schools, Canberra, Australia, February 2–14, 2003, Tübingen, Germany, August 4–16, 2003. Revised Lectures **2004**, 72–112 (2003)
9. Guldogan, E.: Simulation-based analysis for hierarchical storage assignment policies in a container terminal. Simulation **87**(6), 523–537 (2011)
10. Hancock, J.T., Khoshgoftaar, T.M.: Survey on categorical data for neural networks. J Big Data **7**(1), 28 (2020)
11. He, J., Xiao, X., Yu, H., and Zhang, Z.: Dynamic yard allocation for automated container terminal. Ann. Oper. Res. 1–22 (2022)
12. Hun Woo, J., In Cho, Y., Hyeon Yu, S., Hyun Nam, S., Zhu, H., Hoon Kwak, D., and Nam, J.-H.: Machine learning (reinforcement learning)-based steel stock yard planning algorithm. In: 2020 Winter Simulation Conference (WSC), pp. 1560–1571 (2020)
13. Januzaj, Y., Beqiri, E., Luma, A.: Determining the optimal number of clusters using silhouette score as a data mining technique. Int. J. Online Biomed. Eng. (iJOE) **19**, 174–182 (2023)
14. Jolliffe, I.T., Cadima, J.: Principal component analysis: a review and recent developments. Philos. Trans. Royal Soc A: Math., Phys. Eng. Sci. **374**(2065), 20150202 (2016)
15. Kim, K.H., Bae, J.W.: Re-marshaling export containers in port container terminals. Comput. Ind. Eng. **35**(3–4), 655–658 (1998)
16. Kim, K.H., Park, Y.M., Ryu, K.-R.: Deriving decision rules to locate export containers in container yards. Eur. J. Ope. Res. **124**(1), 89–101 (2000)

17. Kuha, J.: AIC and BIC: comparisons of assumptions and performance. Sociol. Methods Res. **33**(2), 188–229 (2004)
18. Legato, P., Canonaco, P., Mazza, R.M.: Yard crane management by simulation and optimisation. Marit. Econ. Logist. **11**(1), 36–57 (2009)
19. Li, M.-K.: Yard storage planning for minimizing handling time of export containers. Flex. Serv. Manuf. J. **27**, 285–299 (2015)
20. Maldonado, S., Gonzalez Ramirez, R., Quijada, F., and Ramirez-Nafarrate, A.: Analytics meets port logistics: a decision support system for container stacking operations. Decis. Support Syst. **121** (2019)
21. Murtagh, F., Contreras, P.: Algorithms for hierarchical clustering: an overview. Wiley Interdisc. Rev.: Data Min. Knowl. Disc. **2**(1), 86–97 (2012)
22. Park, H., Cho, S., Nanda, A., Park, J.: Data-driven dynamic stacking strategy for export containers in container terminals. Flex. Serv. Manuf. J. **35**, 1–26 (2022)
23. Reynolds, D.A., et al.: Gaussian mixture models. Encycl. Biometrics **741**, 659–663 (2009)
24. Schubert, E., Sander, J., Ester, M., Kriegel, H.P., Xu, X.: DBSCAN revisited, revisited: why and how you should (still) use DBSCAN. ACM Trans. Database Syst. (TODS) **42**(3), 1–21 (2017)
25. Sinaga, K.P., Yang, M.-S.: Unsupervised k-means clustering algorithm. IEEE Access **8**, 80716–80727 (2020)
26. Tan, C., He, J., Wang, Y.: Storage yard management based on flexible yard template in container terminal. Adv. Eng. Inform. **34**, 101–113 (2017)
27. Von Luxburg, U.: A tutorial on spectral clustering. Stat. Comput. **17**, 395–416 (2007)
28. Yu, H., Zhang, M., He, J., Tan, C.: Choice of loading clusters in container terminals. Adv. Eng. Inform. **46**, 101190 (2020)
29. Yu, M., Liang, Z., Teng, Y., Zhang, Z., Cong, X.: The inbound container space allocation in the automated container terminals. Expert Syst. Appl. **179**, 115014 (2021)
30. Yuan, C., Yang, H.: Research on k-value selection method of k-means clustering algorithm. **J2**(2), 226–235 (2019)
31. Zhang, C., Liu, J., Wan, Y.-W., Murty, K.G., Linn, R.J.: Storage space allocation in container terminals. Transp. Res. Part B: Methodol. **37**(10), 883–903 (2003)
32. Zhang, C., Wang, Q., Yuan, G.: Novel models and algorithms for location assignment for outbound containers in container terminals. Eur. J. Oper. Res. **308** (2022)

Assessing the Academic Performance of Italian Students Using an Explainable Machine Learning Approach

Stefania Ferrisi

Abstract Educational Data Mining (EDM) is a growing research field, concerned with applying machine learning and data mining techniques to large-scale data from educational settings. One of the main applications of EDM concerns predicting student performance, which can be used to identify at-risk students and define the necessary educational support timely. In this study, a new student performance indicator was proposed relating to the achievement of a degree that combines the three important parameters of degree grade, delay in degree enrollment, and delay in completing a student's university career. Next, a predictive model based on a real history educational dataset collected at the University of Calabria was proposed, applying numerous Machine Learning techniques: Lasso Regressor, Support Vector Machines, Decision Tree, Random Forest, XGBoost, and an Ensemble Model. Results showed that Ensemble Model has the best overall prediction performance. Additionally, explainable AI techniques were applied to analyze the data generating student performance.

Keywords Student performance · Machine learning · Explainable AI

1 Introduction

Educational Data Mining (EDM) is an emerging discipline, concerned with developing methods for exploring the unique and increasingly large-scale data coming from educational settings and using those methods to better understand students' careers and the settings in which they learn in [1]. EDM uses various techniques from statistics and machine learning (ML) to analyze and predict many educational

---

S. Ferrisi (✉)
Departmemt of Mechanical, Energy and Management Engineering, University of Calabria, Rende (CS), Italy
e-mail: stefania.ferrisi@unical.it

M. Barbato et al. (eds.), *Beyond Frontiers of Operations Research: Emerging Technologies and Innovative Optimization Paradigms*, AIRO Springer Series 14,
https://doi.org/10.1007/978-3-031-85894-9_4

aspects such as dropout rate, exam passing rate, student general performance, and so on. In particular, being able to predict students' performance with an appropriate level of accuracy permits providing appropriate feedback on learning processes to both professors and students, which, in turn, could be very important in taking corrective action for at-risk students, or making appropriate adjustments to learning programs and academic regulations. Student success is an important parameter for academic institutions, as it deeply reflects on the quality of the campus and is a relevant requirement for the accreditation of courses and universities. In the paper by Alalawi et al. [2], a systematic review on EDM covering papers from January 2010 to October 2022, the most promising ML approaches were those belonging to ensemble models such as Random Forest (RF) or Extreme Gradient Boosting (XGBoost) [3]. These are ML techniques combining two or more models, overcoming the inherent limitations of single models and obtaining higher accuracy prediction. However, this type of model is considered "black box", meaning that it provides a way to get the result but not to understand how that result was obtained. In the field of education, the complexity and opacity of prediction models make it difficult to identify factors that affect a student's success and the eventual subsequent identification of corrective actions to be taken.

In order to provide data-driven feedback concerning factors that affect students' performance, the use of Explainable Artificial Intelligence (XAI) approaches can be employed. Afzaal et al. [4] developed a dashboard that can help students self-regulate their learning and improve their academic performance based on an explainable RF approach for a programming course. Results demonstrated that the approach used was not only able to identify the potentially influencing factors from courses, but it was also able to identify the necessary change in the feature value to reach high performance.

In this study, various ML techniques combined with an XAI approach were used to predict a student performance indicator. The main contribution consists in developing a new indicator to measure student performance, integrating crucial aspects of the student's university career. These factors are pivotal for companies in candidate selection. Forecasting the indicator at the end of the student's first year of attendance can support the student's efforts to improve their achievement. In addition, the results obtained from XAI tools facilitate the identification of the most influential factors affecting students' careers, thereby enabling the university to implement corrective measures in its course planning. The ML study was developed by exploiting a dataset collected in collaboration with the University of Calabria, ensuring access to high-quality, detailed records. To ensure the suitability of the data for rigorous analysis, a thorough preprocessing phase was necessary, including the merging of multiple datasets and the creation of summary variables to ensure data quality and coherence.

The remainder of this paper is organized as follows: Sect. 2 describes the dataset provided by the University of Calabria and features investigated in the analysis, Sect. 3 describes the ML methods used and Sect. 4 explains the results obtained. Finally, Sect. 5 provides recommendations for future work and gives the conclusion of the study.

2 Data Collection

Data used for the study were provided by the Monitoring, Quality and Evaluation Service, a technical-administrative structure supporting the activities of the Quality Assurance Committee of the University of Calabria (UNICAL), a Southern Italian Campus. Data were derived from a complete career database of all students at UNICAL used, nationally, to monitor and evaluate the quality of teaching in Italian academic institutions. The analysis was conducted on students enrolled between 2013 and 2021. Available data included information on the student's demography (gender, age, residence), previous studies (such as the type of high school and the grade, possible additional completed university careers), student attendance status data (such as attendance year and student regularity), and data on the student's exit from the University (dropout or graduation, degree grade, and exit date). Aiming at defining a student performance indicator that can be calculated at the end of the first year to provide feedback on the performance achieved with respect to future degree acquisition, only information on students enrolled in the first year who had finished or dropped out by the year 2021 were selected.

2.1 *Student Academic Performance*

Student academic performance is the term that measures the achievement of a school title or the completion of a course or test. This term is therefore related to the final grade earned by the student. Relative to the attainment of a bachelor's degree, in order to be a good candidate for a job offer, more and more importance is being given to other factors than the degree grade such as the age of the candidate or the time taken to earn the degree. In this study, a student's performance indicator (SPI) was devised. It combines some items: graduation grade, delayed enrollment in the degree program, and total time to complete the final degree. With the construction of this indicator as described in Eq. (1), the graduation grade earned by a student is commensurate with the other two items. This results in a high performance if the student enrolls in the degree program without delays (e.g., due to course transitions or sabbatical years) and earns the degree in the shortest time possible by obtaining the highest score. As a result, the student who acquires a high performance will be an excellent candidate for the companies to which he or she will apply with an excellent academic background.

$$SPI_i = \frac{VL_i}{VL}\left[\left(1 - \frac{e_i}{e_i + MA_j}\right) + \left(1 - \frac{m_i}{m_i + MT_j}\right)\right] \quad (1)$$

where:

SPI_i = performance indicator for the i_th student;
VL_i = graduation grade earned by the i_th student;
VL = maximum graduation grade equal to the value of 111 (equal to the maximum grade of

110/110 cum laude);
MA_j = minimum age at which the student could enroll in the j_th degree program;
e_i = age of the i_th student at the time of matriculation in the degree program;
MT_j = minimum time required to complete the undergraduate career for the j_th degree program expressed in months;
m_i = time taken by the i_th student to finish the college career expressed in months.

The MA and MT parameters have been identified, for each type of degree course analyzed within the dataset under consideration as, respectively, the age of students most frequently recorded at enrollment in a bachelor's degree and the minimum recorded time taken to graduate. This analysis takes into account the variability in university entry age (17–19 years), which can be attributed to factors such as the month of birth or early entry into elementary school, and the distribution of years spent in university in order to avoid skewing the results. In addition, there has been a nationwide trend in Italy towards a reduction in the average duration of university studies over the years. An evaluation of the final performance indicator reveals that, for the majority of the course, while in the past a student graduating one or two years later was still considered a strong candidate by employers due to the common practice of extending the duration of studies, today this duration is increasingly aligned with the official legal duration of university courses. Finally, in consideration of the aforementioned, the highest possible score was assigned to the two terms comprising the final indicator for students with an associated age or duration value below that of the most frequently observed group.

3 Methods

Before the application of ML algorithms, some necessary pre-processing steps were performed. A data cleaning phase was done to remove or replace missing values and remove duplicate instances. For each student, instances related to passing each exam are present within the dataset. Accordingly, summary variables were constructed to capture certain aspects of a student's career (such as the weighted average of grades earned on passed exams, the number of exams and associated credits earned in the first year). After pre-processing and constructing the variable of performance indicator (as described in Subsect. 2.1), the dataset used for the experiments consisted of a total of 18.387 observations and 23 variables, summarized in Table 1. Descriptive statistics of the variables are displayed in Table 2. Finally, to improve the training and use of ML models, all variables were coded appropriately into numerical forms.

Each student *i* has an associated performance indicator that must be predicted. Since the variable to be predicted is of numerical type, the problem was addressed as a supervised regression problem. The standard workflow for applying ML techniques involves building and tuning the model on a training set and testing its predictive power on a separate test set. In this study, the available dataset was partitioned, with 80% used for training and the most recent 20% for testing. Five different models are employed to predict the performance indicator:

Table 1 Variables description

Description	Name	Definition	Type[a]
Demographic data	Age	Student's age at the time of enrollment	N
	Citizenship	Citizenship of the student (IT for Italian citizenship, FG for foreign citizenship)	C
	Province	Provinces of residence of student	C
	Region	Region of residence of student	C
	Gender	Student gender (female or male)	C
Previous studies data	Num career	Student's progressive career number	N
	HS year	Year in which the student earned the high school degree	C
	HS degree	High school degree grade	N
	HS type	Type of high school (lyceum, technical institute)	C
	Course transition	Equal to 1 if the student has changed degree courses, zero otherwise	C
Attendance status data	Weighted average grades	Weighted average of the grades earned on passing exams	N
	Exam percentage	Number of exams passed compared to the total number of exams in the first year of attendance	N
	Credits percentage	Number of ECTS (associated to exams) passed compared to the total number of ECTS in the first year of attendance	N
	Degree	Type of degree enrolled (bachelor's degree, master's degree)	C
	Study plan	Type of study plan choice by the student (standard or individual)	C
	Course	Name of degree course	C
	Department	Name of institution department to which the course of study belongs	C
	Area	Area of study (STEM, health, humanities)	C
	Tax	Tax band in which the student falls relative to the total fees paid during the year of attendance	C
	Scholarship	Equal to 1 if the student received the scholarship, zero otherwise	C
	Course duration	Course duration in years	C
	Student	Student classification with respect to distance from the university (Off-campus, commuter, on-campus)	C
Exit data	SPI	Student performance indicator	N

[a] N: numeric; C: categoric

- Least Absolute Shrinkage and Selection Operator (LASSO): a regression model using L1 norm for feature selection and regularization to reduce overfitting and increase interpretability;

Table 2 Descriptive statistics of numeric variables

Variable	Min	Mean	Max	Standard dev.
Age	17	21.47	64	3.55
Num career	1	1.55	6	0.64
HS degree	60	86.57	100	11.24
Weighted average grades	18	26.04	33	2.89
Exam percentage	0.04	0.54	1	0.2
Credits percentage	0.03	0.57	1	0.21
SPI	0.4	0.66	1	0.18

- Decision Tree: a heuristic approach that builds a flowchart-like tree structure allowing a clear visualization of the prediction process;
- Random Forest: an ensemble learning approach that combines multiple decision trees to enhance prediction accuracy;
- Extreme Gradient Boost (XGBoost): a boosting algorithm that trains a sequence of weak models (decision trees), each compensating for the weaknesses of its predecessors;
- Support Vector Machines (SVM): a supervised model that tries to find the hyperplane in a high-dimensional space that maximally separates different output values. In this study two different kernel functions are tested: polynomial and RBF (Radial Basis Function).

Finally, an ensemble model was employed. Ensemble learning is a machine learning approach that seeks better predictive performance by combining multiple prediction models, often resulting in higher accuracy compared to individual classifiers in the ensemble [5]. In this work, different combinations of the previously described ML methods were tested. As the combination of RF and XGBoost showed superior predictive performance, only the results of this ensemble model are reported in Sect. 4, as well as only the results of the polynomial SVM, as it showed superior predictive performance compared to other SVM models. All the models have been implemented using R software trained with the optimization algorithms available through the tidyverse package [6]. In detail, a hyperparameter tuning phase was conducted for all five tested methods with the objective of improving their prediction capabilities. The GridSearch with a 5-fold cross-validation method was employed for this phase.

Although ML algorithms show robust predictive power, they act like "black boxe", i.e., they do not allow researchers to understand the process followed by the algorithm to produce the predictions. In response to this critical issue, tools falling under the discipline known as XAI can be adopted to obtain an interpretation of models' results. In this study, both local (i.e., relative to the prediction on a single observation) and global explainability of the models were performed to understand the relationships between independent variables and model output. XAI tools from the DALEX package [7], implemented in the R software, were used.

4 Experimental Results and Discussion

To assess the suggested prediction models, the best-known performance metrics adopted in the case of regression problems were used: root mean square deviation (RMSE), R^2 statistic and mean absolute error (MAE).

RMSE expresses the root of the average amount of deviation of the model response from the true regression plane and depends on the value scale of the target variable. The lower the value taken by this metric, the higher the performance of the model.

$$RMSE = \sqrt{\frac{1}{n}\sum_n (y_i - \hat{y}_j)^2} \quad (2)$$

where:

n is the total number of observations;
y_i is the vector of real values for the target variable;
$\hat{y}_j$ is the vector of predicted values for the target variable.

R^2 statistic is an index, with values between 0 and 1, that measures the proportion of variability in the target variable that can be explained by the independent variables.

$$R^2 = 1 - \frac{\sum_n (y_i - \hat{y}_j)^2}{\sum_n (y_i - \overline{y})^2} \quad (3)$$

where $\overline{y}$ is the average of real values for the target variable.

MAE is the sum of absolute errors divided by the sample size. The smaller the value taken by the metric, the smaller the error made in the prediction.

$$MAE = \frac{1}{n}\sum_n |e_t| \quad (4)$$

where e_t is the vector of errors made in each prediction.

To identify the best model for the problem under consideration, an appropriate trade-off between the values assumed by the evaluation metrics considered must be identified. Table 3 shows the performance achieved by each model.

The best model in terms of the trade-off between evaluation metrics is the RF, followed by the XGBoost with a very close value. The combination of these two models yielded superior prediction performance (as evidenced by the bold values in the table). Regarding the R^2 metric, the resulted values taken by this metric still turns out to be rather low, deviating significantly from the optimal value of 1. This could be

Table 3 Model's performance on test set

	LASSO	DT	SVM	RF	XGBoost	RF+XGBoost
RMSE	0.0699	0.0642	0.0632	**0.0480**	**0.0486**	**0.0454**
R^2 statistic	0.342	0.514	0.642	**0.627**	**0.629**	**0.668**
MAE	0.0571	0.0443	0.0483	**0.0380**	**0.0382**	**0.0353**

due to the type of information provided to the model. However, given the difficulty of universities in obtaining particular student data (such as daily habits, daily study hours, daily course attendance), training a model that is as reliable as possible by considering only data that are easier to collect is preferable. Another aspect related to the low value assumed by the metric under consideration could be that the models were trained for the prediction of student achievement at the end of the first year of attendance. In fact, the first year is very often the most critical one for the university student whose performance can change significantly in subsequent years.

The explainability step was performed on the models identified as best for the problem under investigation: RF and the ensemble RF+XGBoost. In detail, two types of model interpretation were performed: local and global. Regarding the local interpretation, a random observation was extracted from the data set. Then, the Shapley value was used to explain the prediction. The Shapley value, translated to the ML domain, defines the average contribution of each variable to the definition of the final prediction [8]. As shown in Fig. 1, for both models the variable associated with the fraction of credits earned by the student assumes the most significant contribution. Furthermore, although the order changes, most of the remaining variables with higher contributions to prediction do not differ between the two models. Considering the variable "CREDITS PERCENTAGE", Fig. 2 shows how the model response change if the value of this single variable is changed. This type of explanation is called *Ceteris paribus* which is a Latin phrase meaning *other things held constant*. The results show that there is a positive relationship between the examined variable and the target variable, but some differences appear between the two models. Specifically, for both models, if the values taken by the examined variable decrease, the predicted performance indicator also decreases, but a greater impact occurs when considering the ensemble RF+XGBoost. In contrast, if the value of the observed variable increases, there is no impact on the target value for the RF while there is an increase for the RF+XGBoost.

For global explainability, two model interpretation tools are used in this study: the permutation-based variable-importance plot and the analysis of residual. The

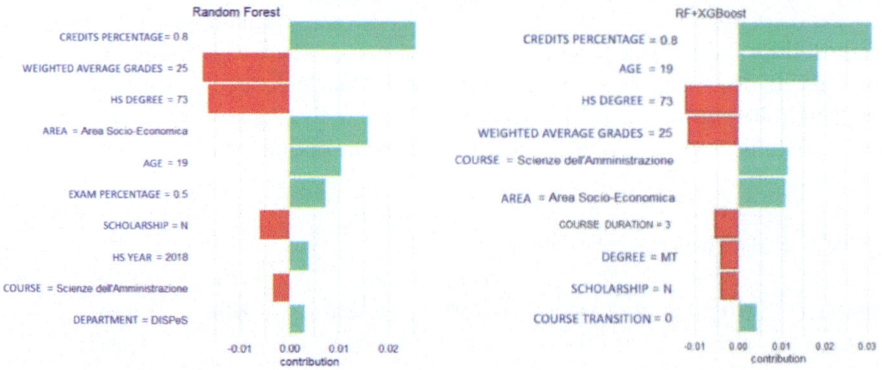

Fig. 1 Shapley values for a random observation with Random Forest and RF+XGBoost

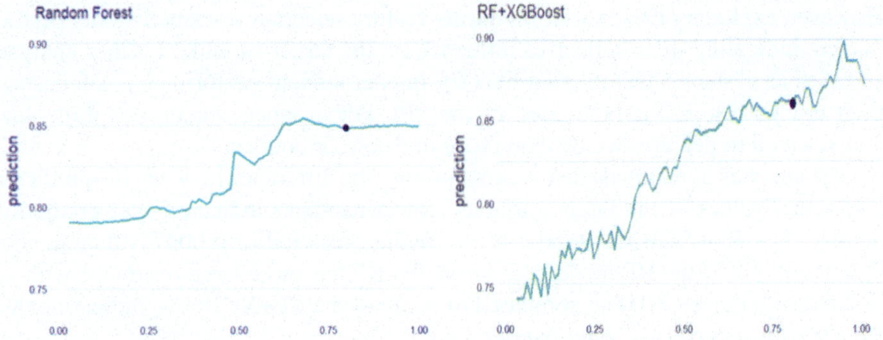

Fig. 2 *Ceteris paribus* for CREDITS PERCENTAGE in a random observation with Random Forest and RF+XGBoost

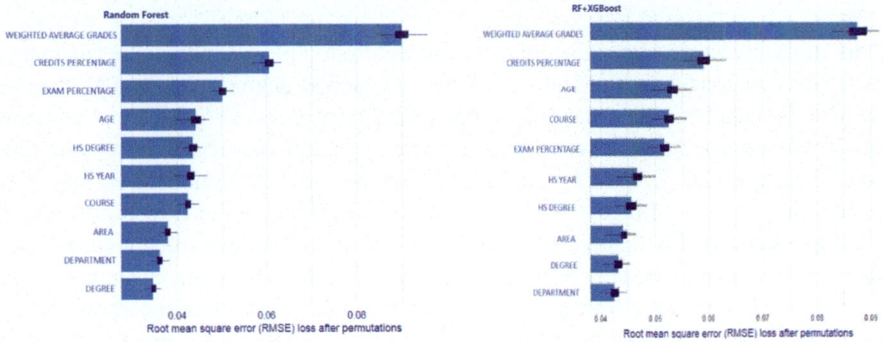

Fig. 3 Variable-importance with 50 permutations for Random Forest and RF+XGBoost

first plot is based on the method described by Fisher et al. [9] in which the main idea is to measure how much the model's performance changes if the effect of variables is removed by generating permutation of the values of the variables. In fact, if a variable is important, after permuting its values, the model's performance will worsen. Considering fifty permutations, as Fig. 3 shows, for both models the two most important features are the same but the weight given to each varies greatly. Considering these top two variables, they are related to the exams passed by the student in the first year of attendance. For the RF model, the next most important feature is "EXAM PERCENTAGE", followed by "AGE" and "HS DEGREE". In contrast, the RF+XGBoost model ranks "AGE" higher than "EXAM PERCENTAGE" and includes "COURSE" among the top five, indicating a shift in feature importance with the combined model. Moreover, although the order varies, the 10 most important variables for prediction turn out to be the same for both models. In detail, seven of these ten variables are related to attendance status, specifically the exams passed by each student. Conversely, variables directly related to the Italian university system, such as course duration or total fees, are not highlighted in Fig. 3.

However, excluding these variables during feature selection worsened model performance, indicating their collective influence on the target variable. Further analysis shows that systemic factors significantly impact student performance, suggesting their crucial role in the prediction of the SPI. For instance, students in three-year courses tend to perform worse than those in five-year courses.

The analysis of residuals involves examining the difference between the predicted and actual values of the target variable. Lower residuals indicate better prediction quality. The RF+XGBoost model has a median residual of 0.0047, often smaller than the 0.009 of the RF model. Additionally, RF has more large residuals, with an interquartile range of 0.099, compared to 0.056 of the RF+XGBoost, indicating that RF+XGBoost performs better overall.

5 Conclusion

This study investigated the formulation and prediction of an academic performance indicator that considers more factors of success such as degree grade and graduation on time. This indicator has the goal of providing a warning for students and realizing academic support early, improving the quality of campus services, and reducing the student failure rate. To identify the best-performing prediction model possible, XAI techniques were combined with ML methods to evaluate better how these models make predictions and identify the model that best fits the college student career data. Results showed that the best-performing model turned out to be the ensemble model RF+XGBoost and that with regard to the variables under consideration, those related to passing exams during the first year of attendance are the most influential for the prediction, closely followed by the demographic variable related to the student's age at the time of enrollment in the degree program. The implementation of this method in a university setting has the potential to significantly impact the academic experience of at-risk students. It provides managers and educators with valuable insights that can be used to inform the development of policies and support systems, thereby fostering a more supportive and effective educational environment. In future, other indicators can be constructed to monitor students, to define a useful dashboard to constantly monitor the students' progress during the frequency of a university course.

References

1. International Educational Data Mining Society. https://educationaldatamining.org
2. Alalawi, K., Athauda, R., Chiong, R.: Contextualizing the current state of research on the use of machine learning for student performance prediction: a systematic literature review. Eng. Rep. **5** (2023). https://doi.org/10.1002/eng2.12699
3. Imran, M., Latif, S., Mehmood, D., Shah, M.S.: Student academic performance prediction using supervised learning techniques. Int. J. Emerg. Technol. Learn. **14**, 92–104 (2019). https://doi.org/10.3991/ijet.v14i14.10310

4. Afzaal, M., Nouri, J., Zia, A., Papapetrou, P., Fors, U., Wu, Y., Li, X., Weegar, R.: Explainable ai for data-driven feedback and intelligent action recommendations to support students self-regulation. Front. Artif. Intell. (2021). https://doi.org/10.3389/frai.2021.723447
5. Optiz, D., Maclin, R.: Popular ensemble methods: an empirical study. J. Artif. Intell. Res. **11** (1999). https://doi.org/10.1613/jair.614
6. Wickham, et al.: Welcome to the Tidyverse. J. Open Sour. Soft. **4** (2019). https://doi.org/10.21105/joss.01686
7. Biecek, P.: DALEX: explainers for complex predictive models in R. J. Mach. Learn. Res. **19**, 1–5 (2018)
8. Štrumbelj, E., Kononenko, I.: An efficient explanation of individual classifications using game theory. J. Mach. Learn. Res. **11**, 1–18 (2010)
9. Fisher, A., Rudin, C., Dominici, F.: All models are wrong, but many are useful: learning a variable's importance by studying an entire class of prediction models simultaneously. J. Mach. Learn. Res. **20**, 1–81 (2019)

Integrating Human-Centric AI in Fashion Industry Optimization: A Step Towards Industry 5.0

Raffaele Dragone, Raffaele Battaglia, and Carmine Cerrone

Abstract This work explores the intersection of advanced manufacturing and the emerging human-centric Industry 5.0 paradigm, specifically within the high-fashion clothing industry. We address the critical issue of optimising cutting operations, a traditionally time-intensive and wasteful process, by minimising setup times and reducing fabric waste. The cutting process primarily consists of laying multiple layers of fabric on a cutting machine and securing templates (stencils) of the parts on top of these layers before the actual cutting can start. We introduce a novel solution that blends the technological advances of Industry 4.0, particularly the integration of mathematical models and the enterprise information system, with the worker-focused ethos of Industry 5.0. The core innovation lies in our use of Large Language Models to translate operators' feedback, expressed in natural language, into quantifiable objectives for linear programming optimisation algorithms. The experimental analysis, conducted on a real-world industrial dataset, validates the effectiveness of the proposed solution, underscoring its potential to enhance industrial operations by harmoniously blending advanced technology with human well-being.

Keywords Human centered AI · Industry 5.0 · Large language model

1 Introduction

The advent of Industry 4.0 marked a significant turning point in the industrial landscape, introducing the integration of industrial machinery with corporate information systems. However, we now stand at the threshold of a new era: Industry 5.0 [1], which emerges as an ambitious journey that redefines the conception and management of industrial processes. While Industry 4.0 focused its attention on digitalisation, Indus-

R. Dragone (✉) · C. Cerrone
Department of Economics and Business Studies, University of Genoa, Genoa, Italy
e-mail: raffaele.dragone@edu.unige.it

R. Battaglia
FourClicks S.r.l. Innovative Startup, Torre Del Greco, Italy

M. Barbato et al. (eds.), *Beyond Frontiers of Operations Research: Emerging Technologies and Innovative Optimization Paradigms*, AIRO Springer Series 14,
https://doi.org/10.1007/978-3-031-85894-9_5

try 5.0 goes a step further by embracing a systemic approach that more deeply fuses digital technologies with the human element. This advancement represents a crucial step towards a human-centric industry, commonly referred to as a human-centred industry 5.0 [2]. In this paper, we focus on the fashion industry, examining a specific case in which a clothing production company aims to transition from the Make-To-Stock (MTS) paradigm to the Make-to-Order (MTO) paradigm [3]. Before the production begins, customer orders for various items to be produced are collected. For each item of clothing, the quantity to be produced is organised by size and colour. One of the most costly, resource-intensive, and time-consuming operations is the preparation and cutting of fabric into different parts of a garment. Initially, an efficient method is needed to cut the various parts of an item, such as the front, back, sleeves, etc., from the cloth for each size. Companies use commercial software to solve this cutting-stock problem [4]. The software constructs outlines indicating how to position the parts on the fabric to maximise the use of raw materials. These outlines, called stencils or (Cutting Patterns), are arranged on the fabric to guide the cutting process. It would be inefficient to cut each stencil separately; therefore, the fabric is laid in multiple layers on a cutting machine. The challenge lies in finding the optimal combination of the number of fabric layers and the set of associated stencils to meet demand while minimising the number of stencils used and fabric consumption. This problem is known as the "Layout Problem," first introduced by Degraeve and Vandebroek [5]. In 2002, Degraeve et al. [6] proposed an alternative formulation of the problem. The fundamental differences in this work compared to the existing literature are twofold. The first is that the sizes contained in each stencil and the set of usable stencils are known a priori. The second difference lies in the objective function, which accounts for both production costs and costs associated with the requests of the cutting operators. These subjective costs, based on operators' observations, are determined using textual reports provided by the operators, interpreted by an AI based on a Large Language Model (LLM) [7].

The motivations behind the development of this new variant of the "Layout Problem" will be introduced in Sect. 2, where we describe the application scenario. In Sect. 3, we present the proposed mathematical model. Section 4 details the experimental results obtained from a test case based on real data. Section 5 presents the conclusions and final observations on this work.

2 Scenario

The necessity that led to the definition of the problem analysed in this work did not arise from a change in the application scenario but rather from a shift in the technological landscape. The application scenario we aim to address is identical to that proposed by Degraeve and Vandebroek [5]. A company receives a series of orders from its customers, these orders pertaining to a specific garment, distinguished by size and colour. The orders are tabulated, with rows corresponding to colours and columns corresponding to sizes. Moreover, the company has a series of stencils, previously developed using commercial software for cutting-stock problems. These stencils are arranged on the fabric to guide the cutting process. Each stencil is associated with a length (the amount of fabric required) and a set of sizes included in the stencil. The cutting process is repeated for each selected stencil. Fabric is laid in multiple layers on a cutting machine, and a stencil is applied on top. The number of garments produced depends on the number of sizes present in the stencil, multiplied by the number of fabric layers. The colours of the garments ultimately depend on the colour of the fabric used in each layer. Therefore, the problem lies in determining which stencils to use and, for each stencil used, how many fabric layers and which colour to spread on the cutting surface. From a technological standpoint, the introduction of automated cutting plans and their integration with the company's information system (Industry 4.0) has accelerated the process. Figure 1a illustrates how starting from the input data, the results obtained from the mathematical model are automatically sent to the cutting machines. These machines, guided by a human operator, can perform their task and provide a set of crucial data for the management of the process. In this scenario, the human operator is purely passive, limited to mechanically performing the tasks assigned. Figure 1b, on the other hand, illustrates the technological solution that we intend to develop in this work. The human-centred paradigm typical of Industry 5.0 has led to the definition of an additional layer of automated analysis. In this layer, the production system's logs, along with operator feedback, are interpreted by an artificial intelligence system based on a Large Language Model (LLM). This is done to include the feedback received as parameters for a subsequent execution of the model. In this scenario, we, therefore, hypothesise a bi-objective model aiming to optimise the functions f_1 and f_2, which respectively describe the operational costs of the system (as seen in existing models in the literature) and the goal of achieving a work process more attentive to the operator's needs. In the centre of the schema shown in Fig. 1b is the operator who intuitively and naturally provides feedback to the system. The interaction between an optimisation model and external feedback is a well-known and utilised approach. An example is the integration of optimisation algorithms and simulation as proposed by Fu [8]. In our case, the feedback is provided by the human operator using natural language. This type of integration between LLM and Optimisation, to the best of the authors' knowledge, represents a novelty in the scientific literature. The role played by the LLM could be replaced by a graphical interface in which the operator may manually adjust the input parameters of the optimisation model. However, this would require specific training for the operator, who would need specific logical and analytical skills for the implemented mathematical model. Using an LLM, the operator is afforded the freedom to effectively contribute to the optimisation process. Furthermore, following their comments in natural language, the operator can view the parameter settings suggested by the LLM and used by the model and consequently understand how their observations have contributed to modifying the setting. This makes the operator part of the process and allows him to acquire more competencies simply by performing their job.

The Industry 4.0 paradigm is primarily focused on system-centric manufacturing. In [2], Lu et al. for Industry 5.0 delineate a hierarchical structure of human needs,

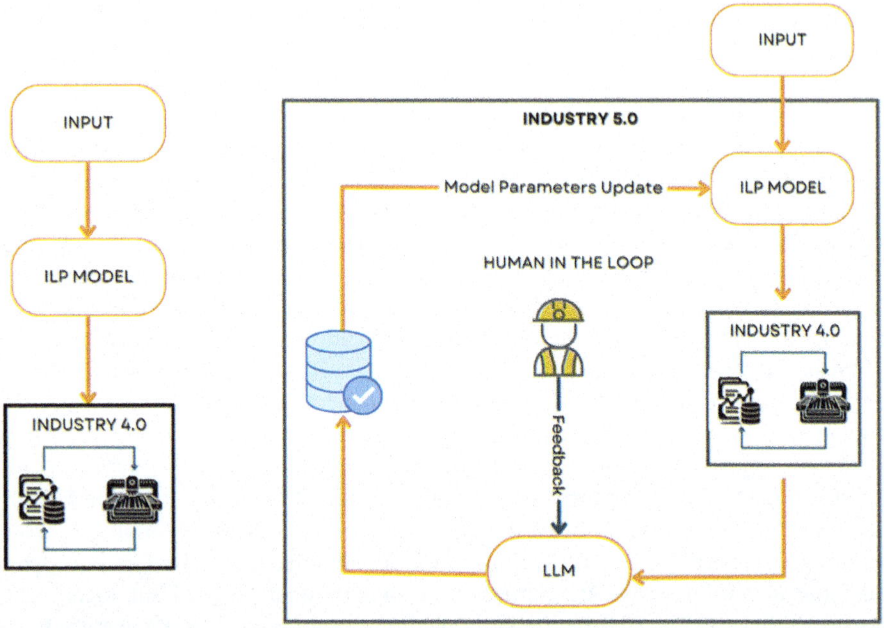

(a) Industry 4.0 (b) Industry 5.0

Fig. 1 Industrial production process flow for Industry 4.0 and 5.0

termed the "Industrial Human Needs Pyramid," which categorises these needs into five distinct levels, ranging from fundamental security to personal growth. The solution implemented in our work for the integration of the optimisation module within a framework based on Industry 5.0 principles succeeds in contributing to achieving levels 2, 3, and 5 of the proposed pyramidal scheme. Level 2 (Health) focusses on creating a working environment that minimises the risk of injury or health damage due to excessive workload. From listening to and interpretation of the operator's requests, there arises directly and immediately the possibility that the mathematical model can reduce the workload if the operator indicates it as too high. Naturally, the model will balance the request to reduce the workload with the need to complete production. Level 3 (Belonging) is focused on collaboration between humans and machines. Our model aims to create a production environment in which humans and machines can establish fruitful relationships and collaborate to achieve the main production goals through communication. Level 5 (Self-actualisation) aspires to enhance workers' self-realisation and personal development. Multiple aspects of the proposed solution contribute to the operator's self-actualisation. The system takes into account the operator's remarks, making them an integral part of the process of improving production. Interaction with an LLM system allows the operator to immediately contribute to the process, even without high-level skills. By this communication, the operator can develop his skills, particularly through the continuous refinement of the interaction with the system.

3 Mathematical Model

The proposed mathematical model requires a comprehensive set of input data to accurately encapsulate the problem. These input data are directly correlated with the specific type of garment being manufactured. The calculation of all associated costs is rigorously based on actual manufacturing expenses. Following this, the model delineates a spectrum of critical parameters that are influenced by the feedback of the operator. These parameters are integral in the development of solutions that deeply respect and value human contribution within the production process. All parameters refer to the estimated cost of the operator effort. These costs are initialised to zero but then enhanced using artificial intelligence as described in the following sections. The parameter $\hat{H}$ represents the maximum number of fabric layers that the operator is willing to manage, while $\hat{\delta}_2$ denotes the cost incurred for each layer that exceeds $\hat{H}$.

Input data:

C : set of colours.

S : set of sizes.

$q(s, c)$: number of garments to be produced per size $s \in S$ and colour $c \in C$

P : set of available Cutting Patterns.

$\hat{q}(p, s)$: number of sizes s occurring in the cutting pattern p.

H : maximum number of layers of cloth for each Cutting Patterns.

ct : cost of fabric per metre.

$l(p)$: length in metres of Cutting Pattern p.

δ_1 : setup cost for Cutting Pattern.

δ_2 : cost of colour setup.

δ_3 : profit associated with overproduced garments.

The cost $\hat{\delta}_1$ represents an additional penalty associated with the need to use an additional Cutting Pattern. This cost is linked to an increase in operator workload. On the other hand, $\hat{\delta}_3$ denotes the cost associated with the requirement to add a fabric of a different colour to the cutting plan, which requires the use of different raw materials and thus complicates the procedure.

The proposed mathematical model is based on four sets of variables. The integer variables x_{pc} indicate, for each Cutting Pattern p, how many layers of colour c will be used. Continuous variables o_p are used in the objective function to calculate the number of excess fabric layers beyond $\hat{H}$ used for each Cutting Pattern p. The binary variables z_{pc} indicate whether the colour c has been used in Cutting Pattern p. These variables are essential because, for each colour used, the corresponding setup cost must be considered. Finally, the variables y_p indicate whether the Cutting Pattern p is used.

Parameters:

$\hat{H}$: prefered number of layers of cloth for each Cutting Patterns.

$\hat{\delta}_1$: setup effort for Cutting Pattern.

$\hat{\delta}_2$: cost of each layer of fabric in excess of the$\hat{H}$th layer.

$\hat{\delta}_3$: cost of color setup.

$$x_{pc} \in \mathbb{Z}^+, \text{ layers of cloth for colour and Cutting Pattern} \qquad \forall p \in P, \forall c \in C$$

$$o_p \in \mathbb{R}^+, \text{ overflow number of layers of cloth respect to} \hat{H} \qquad \forall p \in P$$

$$z_{pc} : \begin{cases} 1, & \text{colour } c \text{ used in Cutting Pattern } p \\ 0 & \text{otherwise} \end{cases} \qquad \forall c \in C, \forall p \in P$$

$$y_p : \begin{cases} 1, & \text{the Cutting Pattern } p \text{ is used} \\ 0 & \text{otherwise} \end{cases} \qquad \forall p \in P$$

To better describe the objective function of the problem, it has been represented as $z = f_1 + f_2$. The function f_1 represents the costs associated with production, with values obtained as input from the problem and depending on the garment to be produced. The function f_2, on the other hand, uses parameters related to feedback derived from Industry 4.0 system logs and human operator feedback. These parameters can evolve over time. Both functions f_1 and f_2 are composed of several components. f_{1a} represents the setup cost of a Cutting Pattern. f_{1b} denotes the cost of the fabric used. f_{1c} corresponds to the setup cost of a colour change. Meanwhile, f_{1d} is the profit associated with the production of garments not present in the order. The profit f_{1d} is related to the multiplier δ_3; When this multiplier is zero, there is no profit derived from excess garment production, while a positive value indicates that excess garments can lead to profit. However, the value of δ_3 should never exceed the cost of fabric associated with garment production. f_{2a} adds an additional cost to the setup of a Cutting Pattern derived from operational feedback. f_{2b} is the cost associated with each layer of fabric that exceeds $\hat{H}$. f_{2c} is the cost of setting up a new colour on the machine. All costs f_2 are valued on the basis of feedback received from system logs and human operators. These costs will always be analytically highlighted in the solutions produced.

$$f_{1a} = \delta_1 \sum_{p \in P} y_p \tag{1}$$

$$f_{1b} = \sum_{p \in P} \sum_{c \in C} l(p) x_{pc} ct \tag{2}$$

$$f_{1c} = \delta_2 \sum_{p \in P} \sum_{c \in C} z_{pc} \tag{3}$$

$$f_{1d} = \delta_3 \sum_{s \in S} \sum_{c \in C} \left[\left(\sum_{p \in P} x_{pc} \hat{q}(p, s) \right) - q(s, c) \right] \tag{4}$$

$$f_{2a} = \hat{\delta}_1 \sum_{p \in P} y_p \tag{5}$$

$$f_{2b} = \hat{\delta}_2 \sum_{p \in P} o_p \tag{6}$$

$$f_{2c} = \hat{\delta}_3 \sum_{p \in P} \sum_{c \in C} z_{pc} \tag{7}$$

$$f_1 = f_{1a} + f_{1b} + f_{1c} - f_{1d} \tag{8}$$

$$f_2 = f_{2a} + f_{2b} + f_{2c} \tag{9}$$

To solve the problem, we have defined a Mixed Integer Linear Programming Model capable of generating the optimal solution. The objective function z, which is to be minimised, has been divided into components (1)–(7).

$$\min \quad z = f_1 + f_2 \tag{10}$$

s.t.

$$\sum_{p \in P} \hat{q}(p, s) x_{pc} \geq q(s, c) \qquad \forall s \in S, \forall c \in C \tag{11}$$

$$\sum_{c \in C} x_{pc} \leq \hat{H} + o_p \qquad \forall p \in P \tag{12}$$

$$x_{pc} \leq H z_{pc} \qquad \forall p \in P, \forall c \in C \tag{13}$$

$$z_{pc} \leq y_p \qquad \forall p \in P, \forall c \in C \tag{14}$$

$$o_p \in [0, H - \hat{H}] \qquad \forall p \in P \tag{15}$$

$$y_p \in \{0, 1\} \qquad \forall p \in P \tag{16}$$

$$z_{pc} \in \{0, 1\} \qquad \forall p \in P, \forall c \in C \tag{17}$$

$$x_{pc} \in \mathbb{Z}^+ \qquad \forall p \in P, \forall c \in C \tag{18}$$

The constraint family (11) ensures that a sufficient number of garments are produced to meet all orders for each size $s \in S$ and colour $c \in C$. The constraint family (12) guarantees for each Cutting Pattern $p \in P$ that the maximum number of fabric

Fig. 2 Test garment

layers H is not exceeded while also allowing the counting of excess layers beyond $\hat{H}$ using the variable o_p. The constraint family (13) ensures that fabric layers of a specific colour x_{pc} can only be created if that colour is active for the given Cutting Pattern z_{pc}. The constraint family (14) serves to prevent the use of any colour for the Cutting Pattern $p \in P$ if p is not active.

4 Computational Experiments

An experimental test was conducted using real-world data to validate the proposed Mathematical Model. A company leader in the fashion sector kindly provided these data. All data are related to the garment shown in Fig. 2. Table 1 shows the production quantities, organised by size and colour.

Figure 3 shows an example of a Cutting Pattern that includes a garment of each size from XS to XXL. Table 2, on the other hand, presents information about all cutting

Table 1 Garments ordered by customers and to be produced by size and colour

C/S	XS	S	M	L	XL	XXL
Green	13	17	15	18	16	12
Violet	14	17	18	15	10	4
Black	34	39	44	38	36	32
White	54	59	62	56	46	44

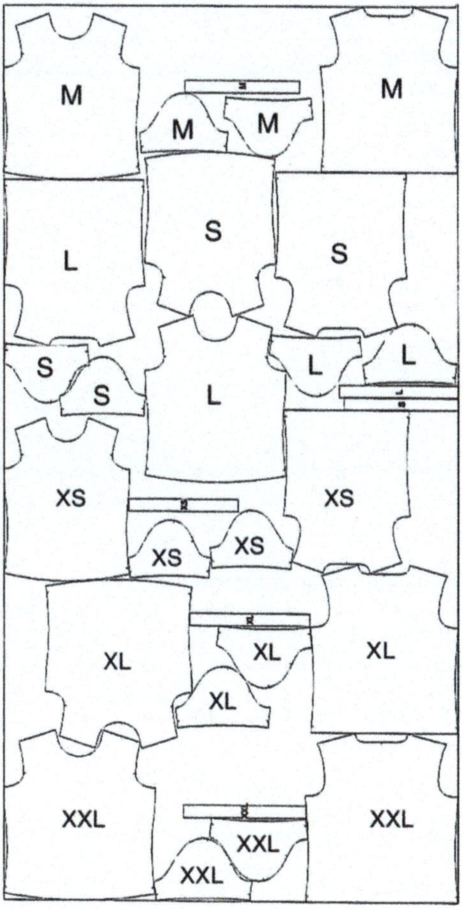

Fig. 3 Cutting Pattern 2

patterns provided. Specifically, for each Cutting Pattern, the fabric consumption in metres and the sizes included are specified.

For the garment, we have a cost of $ct = 7.2$ € per metre of fabric. The setup cost for each Cutting Pattern is $\delta_1 = 100$ €. The setup cost for changing colour is $\delta_2 = 25$ €. The potential profit from each excess garment produced is set at $\delta_3 = 0.5$ €.

The code was implemented in Java and CPLEX 22.11. The tests were carried out on a MacBook Air with an M2 chip and 8GB of RAM.

Tables 3, 4 show the results obtained by varying the parameters $\hat{\delta_1}$ and $\hat{\delta_2}$. The column $\#OP$ indicates the number of garments cut in excess of the requirements stated in Table 1. The column $\#PT$ shows the number of Cutting Patterns used in the solution. The column $\#CL$ shows the number of different colours used in all Cutting Patterns. The column $\#OL$ displays the total of the overflow layers with respect to the parameter $\hat{H}$. The first row of Table 3 is crucial as it illustrates the solution produced

Table 2 Cutting patterns

		Sizes					
	Length (Mt.)	XS	S	M	L	XL	XXL
Pattern 1	**3.44**	3				3	1
Pattern 2	**3.23**	1	1	1	1	1	1
Pattern 3	**3.17**		2	2	2		
Pattern 4	**1.63**		1	1	1		
Pattern 5	**1.41**					1	1
Pattern 6	**0.73**	1					
Pattern 7	**0.74**		1				
Pattern 8	**0.76**			1			
Pattern 9	**0.78**				1		
Pattern 10	**0.79**					1	
Pattern 11	**0.82**						1

Table 3 Results obtained by varying the $\hat{\delta_1}$ parameter

$\hat{\delta_1}$	f_1	f_2	f_{1a}	f_{1b}	f_{1c}	f_{1d}	f_{2a}	f_{2b}	f_{2c}	#OP	#PT	#CL	#OL
0	3552	0	400	2983	200	31	0	0	0	62	4	8	18
20	3552	80	400	2983	200	31	80	0	0	62	4	8	18
40	3552	160	400	2983	200	31	160	0	0	62	4	8	18
60	3552	240	400	2983	200	31	240	0	0	62	4	8	18
80	3552	320	400	2983	200	31	320	0	0	62	4	8	18
100	3638	300	300	3150	250	62	300	0	0	124	3	10	21

in the absence of any feedback received, thus setting the parameters $\hat{\delta_1}$, $\hat{\delta_2}$, and $\hat{\delta_3}$ to zero. In this scenario, we have $z = f_1 = 3552$ €, representing the minimum cost to execute the chosen cut using the available Cutting Patterns. Here, we observe $\#OP = 62$, indicating that 62 additional pieces were produced beyond the required amount, using 4 Cutting Patterns and a total of 8 colour changes. It is noticeable that the fabric cost f_{1b} contributes 2983 €, while the costs associated with the cutting process are $f_{1a} + f_{1c} = 600$ €. Another important insight that we can obtain from Table 3 is that, as $\hat{\delta_1}$ varies, the number of Cutting Patterns used remains constant at 4, thus not altering the production cost f_1. Only by employing $\hat{\delta_1} = 100$ do we achieve $\#PT = 3$, saving one Cutting Pattern, which leads to a reduction of 100 € in f_{1a}, but an increase in the fabric cost f_{1b} by 167 €. Table 4 reveals that as the cost $\hat{\delta_2}$ associated with excess layers beyond the parameter $\hat{H}$ increases, we can achieve solutions with $\#OL = 0$. However, this leads to an increment by one in the number of Cutting Patterns used and, consequently, to an increase in production costs.

Table 4 Results obtained by varying the $\hat{\delta_2}$ parameter

$\hat{\delta_2}$	f_1	f_2	f_{1a}	f_{1b}	f_{1c}	f_{1d}	f_{2a}	f_{2b}	f_{2c}	#OP	#PT	#CL	#OL
2	3558	30	400	2962	225	29	0	30	0	58	4	9	15
4	3568	44	400	2917	275	24	0	44	0	48	4	11	11
6	3568	66	400	2917	275	24	0	66	0	48	4	11	11
8	3568	88	400	2917	275	24	0	88	0	48	4	11	11
10	3675	0	500	2924	275	24	0	0	0	48	5	11	0
12	3675	0	500	2924	275	24	0	0	0	48	5	11	0

4.1 *Industry 5.0 Cycle*

In this section, we present the results of an experiment where we tested the scheme depicted in Fig. 1b. Essentially, starting from the results obtained in the first row of Table 3, we analyse the behaviour of the solution generated by the model using the parameters received from the operator. After each solution produced by the mathematical model, we report the feedback provided by the operator and their translation into parameters using a Large Language Model. This process will be iterated to analyse the resultant outcomes. The Large Language Model used is ChatGPT-4 [9]. An online GPTs [10] has been customised to execute this task. Table 5 presents a demonstration of the execution of the process shown in Fig. 1b. The row *Run*1 reports the results described in Table 3, which can be used as a starting point. Subsequently, we received the first feedback, leading to an increase of $\hat{\delta_2} = 4$ in the *Run*2 row.

This row shows a cost increase for f_1 of 16 €, but a desired decrease of 8 layers. The second feedback leads to $\hat{\delta_2} = 8$, which, however, does not alter the optimal

Table 5 Results obtained by interaction with LLM

	$\hat{\delta_1}$	$\hat{\delta_2}$	$\hat{\delta_3}$	f_1	f_2	#OP	#PT	#CL	#OL
Run 1	0	0	0	3552	0	62	4	8	18
Feedback 1	*Use fewer fabric layers*								
Run 2	0	4	0	3568	44	48	4	11	11
Feedback 2	*There are still too many fabric layers*								
Run 3	0	8	0	3568	88	48	4	11	11
Feedback 3	*There are too many fabric layers*								
Run 4	0	12	0	3675	0	48	5	11	0
Feedback 4	*I have loaded too many rolls of fabric*								
Run 5	0	12	20	3684	200	56	5	10	0
Feedback 5	*I have used too many Cutting Patterns*								
Run 6	25	12	20	3684	325	56	5	10	0
Feedback 6	*I prefer to use fewer Cutting Patterns*								
Run 7	50	12	20	3558	560	58	4	9	15

solution. The third feedback, which results in $\hat{\delta_2} = 12$, leads to a further increase in the cost of f_1 of 7 €, but reduces the excess layers to zero. After incorporating all the 6 feedback, we obtain a parameter setting of $\hat{\delta_1} = 50, \hat{\delta_2} = 12, \hat{\delta_3} = 20$, which results in a negligible cost increase of only 6 € compared to the initial solution, returning to the use of 4 Cutting Patterns but again increasing the number of excess layers. This test demonstrates how the system can respond to operator requests while never straying too far from the optimal solution. Moreover, allowing the operator to control the process enables them, iteration after iteration, to understand the optimisation process employed. The final solution is very similar to the initial one, but the process used has engaged the operator, allowing them to fully comprehend the implications associated with various optimisation criteria.

5 Conclusions

This study represents a notable advancement in aligning technological progress with the human-centric philosophy of Industry 5.0, especially within the high-fashion clothing industry. Our methodology, which merges the accuracy of mathematical models and enterprise information systems from Industry 4.0 with the empathetic, worker-centric approach of Industry 5.0, has shown encouraging results in optimising cutting processes. This is achieved by incorporating Large Language Models to convert operators' natural language feedback into parameters for a linear programming optimisation model. The encouraging results of our empirical analysis, using actual industrial data, suggest the practical viability of our method in real industrial contexts. However, future efforts should focus on broadening the experimental phase to gain a more thorough understanding of the impact of the approach. A particularly intriguing area for future research is the collection and examination of data from the real-world deployment of this Industry 5.0 project to further evaluate this methodology's contribution. Moreover, a noteworthy development in subsequent investigations would be to enhance the mathematical model, especially concerning utilising a finite number of Cutting Patterns. The existing limitation could be addressed by increasing the number of patterns or devising a dynamic method for generating Cutting Patterns. In summary, this research adds value to the field by offering an innovative solution to a long-standing industrial problem and laying the groundwork for future studies that could further exploit the synergy of advanced technology and human feedback, a hallmark of Industry 5.0.

Acknowledgements This work was carried out within the framework of the project "RAISE—Robotics and AI for Socio-economic Empowerment" and has been supported by European Union—NextGenerationEU.

References

1. Breque, M., De Nul, L., Petridis, A.: European Commission, and Directorate-General for Research and Innovation, Industry 5.0: towards a sustainable, human-centric and resilient European industry, Publications Office (2021)
2. Lu, Y., Zheng, H., Chand, S., Xia, W., Liu, Z., Xu, X., Bao, J.: Outlook on human-centric manufacturing towards Industry 5.0. J. Manuf. Syst. **62**, 612–627 (2022)
3. Rajagopalan, S.: Make to order or make to stock: model and application. Manag. Sci. **48**(2), 241–256 (2002)
4. Farley, A.A.: Mathematical programming models for cutting-stock problems in the clothing industry. J. Oper. Res. Soc. **39**(1), 41–53 (1988)
5. Degraeve, Z., Vandebroek, M.: A mixed integer programming model for solving a layout problem in the fashion industry. Manag. Sci. **44**(3), 301–310 (1998)
6. Degraeve, Z., Gochet, W., Jans, R.: Alternative formulations for a layout problem in the fashion industry. Eur. J. Oper. Res. **143**(1), 80–93 (2002)
7. Chen, W., Yan-yi, L., Tie-Zheng, G., Da-Peng, L., Tao, H., Zhi, L., Ying-You, W.: Systems engineering issues for industry applications of large language model. Appl. Soft Comput. **151**, 111165 (2024)
8. Fu, M.C.: Optimization for simulation: theory vs. practice. INFORMS J. Comput. **14**(3), 192–215 (2002)
9. OpenAI: ChatGPT-4 (2023). https://openai.com/chatgpt-4
10. Human Centered AI for Industry 5.0 GPTs. https://chat.openai.com/g/g-acceLtfWF-human-centered-ai-for-industry-5-0

Operations Research Applications to Contemporary Optimization Challenges

Column Generation Algorithms for the Pickup and Delivery Problem with Time Windows and Last-in-First-out Loading

Dario Palasgo, Matteo Fischetti, and Roberto Roberti

Abstract We study the *Pickup and Delivery Problem with Time Windows and Last-in-First-out Loading* (PDPTWL), a decision-making problem that aims at minimizing the cost to serve a set of customers (consisting of pickup and delivery locations) within their time windows, using a fleet of capacitated vehicles and handling their loads with a Last-in-First-out policy. We propose a bounding procedure based on column generation to find tight dual bounds to the PDPTWL by solving the linear relaxation of a set partitioning formulation, where variables correspond to (non-necessarily elementary) routes. We consider a set of benchmark instances from the literature to show that these dual bounds are tight and can be computed in a few seconds. Therefore, the bounding procedure can be a building block of an exact method for the PDPTWL.

Keywords Pickup and delivery · Time windows · LIFO constraints · Vehicle routing · Column generation · Dynamic programming

1 Introduction

The *Pickup and Delivery Problem with Time Windows and Last-in-First-out Loading* (PDPTWL) is a one-to-one pickup and delivery vehicle routing problem with time windows (see [9, 12, 13]) with the additional constraint that commodities are unloaded in the *Last-in-First-out* (LIFO) sequence. The PDPTWL arises in distribution settings where loads are challenging to handle because specific equipment is needed and loading/unloading operations are time-consuming or dangerous. This

D. Palasgo (✉) · M. Fischetti · R. Roberti
Department of Information Engineering, University of Padova, Padova, Italy
e-mail: dario.palasgo@phd.unipd.it

M. Fischetti
e-mail: matteo.fischetti@unipd.it

R. Roberti
e-mail: roberto.roberti@unipd.it

M. Barbato et al. (eds.), *Beyond Frontiers of Operations Research: Emerging Technologies and Innovative Optimization Paradigms*, AIRO Springer Series 14,
https://doi.org/10.1007/978-3-031-85894-9_6

is the case, for example, when transporting animals, hazardous materials, or heavy goods, where a LIFO policy to load/unload vehicles is recommended or even mandatory. LIFO constraints limit the freedom in the decision-making process of designing the routes because the sequence of pickups of each route determines the exact order of the visits to the corresponding deliveries.

The PDPTWL can be described as follows. A directed (loop-free) graph $G = (V, A)$ is given. The vertex set V is defined as $V = N \cup \{0\} = P \cup D \cup \{0\}$, where $N = P \cup D$ is the set of customer locations (or simply customers), $P = \{1, 2, \ldots n\}$ is a set of n pickup locations, $D = \{n+1, n+2, \ldots, 2n\}$ is a set of n delivery locations, and 0 represents a depot where an unlimited fleet of vehicles, each one with capacity Q, is located. The customer locations are paired in n requests, each consisting of a pickup location $i \in P$ and a corresponding delivery location $n+i \in D$. Satisfying customer request $i = 1, \ldots, n$ requires moving some commodity of load $q_i > 0$ from pickup $i \in P$ to delivery $n+i \in D$ (where $q_{n+i} = -q_i$); to simplify the presentation, a dummy quantity $q_0 = 0$ is associated with the depot. A time window $[e_0, l_0]$ is associated with the depot to represent the vehicle's working horizon. Each customer $i \in N$ is associated with a time window $[e_i, l_i]$, representing the time when vehicles can serve the customer. Time windows are *hard*, i.e., vehicles can visit customer i at any time no later than l_i, but the service cannot start before e_i. The arc set is defined as $A = \{(0, j) \mid j \in P\} \cup \{(i, 0) \mid i \in D\} \cup \{(i, j) \mid i \in P, j \in N \backslash \{i\} : e_i + t_{ij} \leq l_j, q_i + q_j \leq Q\} \cup \{(i, j) \mid i \in D, j \in N \backslash \{i, i-n\}, e_i + t_{ij} \leq l_j, |q_i + q_j| \leq Q\}$, where $t_{ij} > 0$ is the travel time to traverse arc $(i, j) \in A$, which includes the time to serve i (if i is a customer). A travel cost $c_{ij} > 0$ is also associated with arc $(i, j) \in A$, and a fixed cost C is incurred for each vehicle deployed. As commonly done in the literature (see, e.g., [1]), travel costs and travel times satisfy the triangular inequality. Without loss of generality, we also assume that $e_i \geq e_0 + t_{0i}$ for each pickup $i \in P$, and $l_i \leq l_0 - t_{i0}$ for each delivery $i \in D$. A vehicle route is feasible if it starts and ends at the depot, visits a set of customers within their time windows without exceeding the vehicle capacity, and fulfills the LIFO constraints (which implies that each pair of pickup and delivery customers is assigned to the same vehicle route and the pickup is visited before the delivery). The objective of the PDPTWL is to find a set of routes that satisfy all customer requests and minimize the total costs (travel costs plus fixed costs).

The PDPTWL was first introduced in [3], where three exact branch-price-and-cut algorithms managing the LIFO constraints in different ways are presented. The state-of-the-art exact method for the PDPTWL is provided in [1], where a relaxed network flow model with side constraints is used; this network flow model is based on the concept of *fragments*, i.e., feasible paths starting with an empty vehicle from a pickup, ending at the corresponding delivery with the empty vehicle, and serving, in between, some customers within their time windows without violating capacity constraints while adhering with the LIFO constraint. The idea of fragments was the key ingredient of two other exact methods later developed for problems related to the PDPTWL, i.e., a branch-and-cut algorithm for the dial-a-ride problem [15] and a column and Benders-cut-constraints generation method for the selective dial-a-ride [14]. Two notable metaheuristics for the PDPTWL are proposed in [4, 11].

The main contribution of our work is to provide a bounding procedure, based on column generation, to provide tight lower bounds to the PDPTWL by solving the linear relaxation of a set partitioning formulation strengthened with some valid inequalities. The pricing algorithm relies upon a labeling algorithm based on the fragments. To lower the complexity of the pricing problem, we allow routes to be non-elementary, but we prevent some cycles through partial elementarity. To speed up the bounding procedure, we properly stabilize the master problem. Through a computational campaign on benchmark instances from the literature, we show that the proposed bounding procedure provides high-quality lower bounds to the PDPTWL in a few seconds of computing time. This can be a promising building block of an exact algorithm for the PDPTWL.

The remainder of this paper is organized as follows. Section 2 describes a *Mixed-Integer Linear Programming* (MILP) formulation of the PDPTWL based on the fragments, which features a polynomial number of constraints. Section 3 presents a set-partitioning formulation for the PDPTWL. Section 4 first describes the proposed bounding procedure at a high-level and then summarizes its main components. Section 5 illustrates the computational results achieved with the bounding procedure. Finally, some conclusions are drawn in Sect. 6.

2 Fragment-Based Mathematical Model

In this section, we formally introduce the concept of fragment and provide a fragment-based MILP model for the PDPTWL.

A fragment is a directed elementary path that (i) starts at a pickup with an empty vehicle within its time window, (ii) ends at the corresponding delivery with an empty vehicle within its time window, (iii) visits a set of customers in between by satisfying the capacity, time-window, and LIFO constraints. We can observe that any feasible PDPTWL route can be seen as a sequence of fragments connected by arcs linking either the depot to a pickup, or a delivery to a pickup, or a delivery to the depot; the vehicle is empty while traversing such arcs. Indeed, consider the following route $r = (0 - 1 - 2 - 12 - 11 - 3 - 4 - 14 - 5 - 15 - 13 - 0)$, where $n = 10$, so $P = \{1, 2, \ldots, 10\}$, and $D = \{11, 12, \ldots, 20\}$. Let us assume that the route is feasible. We can see that route r is made up of five components, i.e., arc $(0, 1)$, fragment $1 - 2 - 12 - 11$, arc $(11, 3)$, fragment $3 - 4 - 14 - 5 - 15 - 13$, and arc $(13, 0)$. Arcs $(0, 1)$, $(11, 3)$, and $(13, 0)$ are traversed with an empty vehicle, and the two fragments embed the capacity and LIFO constraints.

To describe our fragment-based MILP model, we need some additional notation. Let Φ be the set of undominated fragments, and let $\Phi_i \subseteq \Phi$ be the set of undominated fragments starting from pickup $i \in P$. For each fragment $f \in \Phi$, let t_f be the minimum travel time (including necessary waiting times) of the fragment, and let e_f and l_f be the first and last departure time of the fragment from its starting pickup so that the travel time to reach the last delivery location is t_f and the fragment maintains its feasibility. We denote by $[e_f, l_f]$ the time window of fragment $f \in \Phi$. A cost

$c_f > 0$, computed as the sum of the costs of its arcs, is associated with each fragment $f \in \Phi$. For each fragment $f \in \Phi$ and each customer $i \in N$, let a_{fi} be a binary coefficient equal to 1 if fragment f serves customer i (0 otherwise). Set Φ does not contain dominated fragments, which means that, for each fragment $f \in \Phi$, Φ does not contain any other fragment g such that $f, g \in \Phi_i$, $c_g \leq c_f$, $t_g \leq t_f$, $l_g \geq l_f$, $e_g + t_g \leq e_f + t_f$, and $a_{fi} = a_{gi}$, for each $i \in N$. For conciseness, we omit the proof of this dominance rule.

We also define the set $E \subset A$ of linking arcs as $E = E^{0P} \cup E^{DP} \cup E^{D0}$, where $E^{0P} = \{(0, j) \mid j \in P\}$, $E^{D0} = \{(i, 0) \mid i \in D\}$, and $E^{DP} = \{(i, j) \mid i \in D, j \in P \setminus \{i - n\}, e_i + t_{ij} \leq l_j\}$.

In the following, we use two symbols ($\prec$ and $\not\prec$) to indicate the precedence relationships between pairs of customers. If customer $i \in N$ must be visited earlier than customer $j \in N$ in any solution, we say that $i \prec j$—this happens, e.g., if $l_i + t_{ij} \leq e_j$. For a pair of customers $i, j \in N$, if we cannot say whether $i \prec j$ or $j \prec i$, we write $i \not\prec j$.

The fragment-based MILP model uses three sets of variables: $y_f \in \{0, 1\}$, a binary variable equal to 1 if fragment $f \in \Phi$ is traversed by any vehicle (0 otherwise); $x_{ij} \in \{0, 1\}$, a binary variable equal to 1 if linking arc $(i, j) \in E$ is traversed by a vehicle (0 otherwise), and $\beta_i \in \mathbb{R}_+$, an auxiliary non-negative real variable representing the time at which customer $i \in N$ is served.

The PDPTWL can be formulated as follows:

$$\min_{\substack{y \in \{0,1\}^{|\Phi|} \\ x \in \{0,1\}^{|E|} \\ \beta \in \mathbb{R}_+^{|N|}}} \sum_{f \in \Phi} c_f y_f + \sum_{(i,j) \in E^{DP} \cup E^{D0}} c_{ij} x_{ij} + $$

$$+ \sum_{(0,j) \in E^{0P}} (C + c_{ij}) x_{0j} \tag{1a}$$

$$\text{s.t.} \sum_{f \in \Phi} a_{fi} y_f = 1 \qquad \forall i \in P \tag{1b}$$

$$\sum_{(i,j) \in E^{0P} \cup E^{DP}} x_{ij} = \sum_{f \in \Phi_j} y_f \qquad \forall j \in P \tag{1c}$$

$$\sum_{f \in \Phi_{(i-n)}} y_f = \sum_{(i,j) \in E^{DP} \cup E^{D0}} x_{ij} \qquad \forall i \in D \tag{1d}$$

$$\beta_{(i+n)} \geq \beta_i + \sum_{f \in \Phi_i} t_f y_f \qquad \forall i \in P \tag{1e}$$

$$\beta_i \geq e_i + \sum_{f \in \Phi_i} (e_f - e_i) y_f \qquad \forall i \in P \tag{1f}$$

$$\beta_i \leq l_i + \sum_{f \in \Phi_i} (l_f - l_i) y_f \qquad \forall i \in P \tag{1g}$$

$$\beta_j \geq \beta_i + t_{ij} x_{ij} \qquad \forall (i, j) \in E^{DP} : i \prec j \tag{1h}$$

$$\beta_j \geq \beta_i + t_{ij} - M(1 - x_{ij}) \qquad \forall (i, j) \in E^{DP} : i \not\prec j \tag{1i}$$

The objective function (1a) aims at minimizing the total variable and fixed costs. Constraints (1b) guarantee that each pickup (and consequently each delivery) is visited exactly once. Constraints (1c) ensure the flow conservation at pickups whereas constraints (1d) ensure flow conservation at deliveries. Constraints (1e) align the visit time of the starting pickup of a fragment with the visit time of the corresponding delivery at the end of the fragment if such a fragment is selected. Constraints (1f) and (1g) force the visit time of the initial pickup node of a fragment to fulfill the corresponding fragment's time window. Constraints (1h) and (1i) play the same role as (1e) and define coherent time visits of the endpoints of the linking arcs—in (1i), M is a sufficiently large coefficient.

Notice that LIFO and capacity constraints are modeled through the feasibility of fragments, and routes are obtained as sequences of fragments connected by linking arcs. Moreover, formulation (1) has an exponential number of variables and a polynomial number of constraints. If the entire set of fragments can be enumerated, formulation (1) can be solved by using a black-box MILP solver. Otherwise, column-generation techniques are necessary.

3 Set Partitioning Model

The PDPTWL can also be formulated as a set partitioning problem to obtain a model whose linear relaxation is tighter than the linear relaxation of model (1).

Let us define the set R of all elementary routes starting and ending at the depot, built as sequences of fragments and linking arcs, satisfying the time-window constraints. Each route $r \in R$ has a cost c_r equal to the sum of the costs of the corresponding fragments and linking arcs. Let a_{ri} be a binary coefficient equal to the number of times route r visits pickup $i \in P$. By introducing a binary variable $z_r \in \{0, 1\}$ equal to 1 if route $r \in R$ is traversed by a vehicle (0 otherwise), a set partitioning model for the PDPTWL is:

$$z^* = \min_{z \in \{0,1\}^{|R|}} \sum_{r \in R} (C + c_r) z_r \tag{2a}$$

$$\text{s.t.} \quad \sum_{r \in R} a_{ri} z_r = 1 \qquad \forall i \in P \tag{2b}$$

The objective function (2a) minimizes the fixed and variable cost of the traversed routes, and constraints (2b) ensure that each customer request is satisfied exactly once. The number of routes of the set R grows exponentially with the number of customers, so formulation (2) can be used to compute lower bounds or optimal solutions to the PDPTWL only through column generation. The bounding procedure we provide is based on formulation (2) and column generation. We refer the reader to [6] for an introduction to column generation.

4 The Lower Bounding Procedure

The lower bounding procedure we propose features three main steps:

1. *Enumerate the set of fragments*—The first step consists of enumerating the set of undominated fragments Φ. We deploy the depth-first search enumeration algorithm described in [1], where all details can be found. As it was possible to enumerate all fragments on all test instances from the literature, we have not considered the case the set Φ is too large to enumerate. However, it could still be possible to use the lower bounding procedure, even without having the entire set Φ, by adjusting the pricing problem in the next steps and considering a delayed column generation procedure for the set Φ.
2. *Compute a lower bound to the number of vehicles*—The second step aims at finding a tight estimate (that we call nv) of the number of vehicles required to serve all customers. To do so, we solve the following linear problem derived from formulation (2) by changing the objective function and relaxing integrality constraints:
$$nv = \min_{z \in \mathbb{R}_+^{|R|}} \left\{ \sum_{r \in R} z_r \mid \sum_{r \in R} a_{ri} z_r = 1, \forall i \in P \right\} \tag{3}$$
To compute nv, we rely on column generation. The pricing problem of finding negative reduced-cost routes is an *Elementary Shortest Path Problem with Resource Constraints* (ESPPRC), known for being NP-hard (see [8]), which is commonly solved with labeling algorithms. Instead of generating elementary routes, we replace the set R with a set $T \supset R$ of non-necessarily elementary routes where a limited number of subtours is cut off so that the pricing problem is computationally faster to solve. In particular, we devise a labeling algorithm that combines two known techniques from the literature: the t-route state space relaxation (introduced in [5]), and the partial elementarity (introduced in [7]) that prevents multiple visits to a set of selected customers in each route. To accelerate convergence, we stabilize the dual problem with an in-out technique [2], see Sect. 4.3.
3. *Compute a lower bound to the PDPTWL*—The last step consists of computing the following lower bound, lb, to PDPTWL:
$$lb = \min_{z \in \mathbb{R}_+^{|R|}} \left\{ \sum_{r \in R} (C + c_r) z_r \mid \sum_{r \in R} a_{ri} z_r = 1, \forall i \in P, \text{ and } \sum_{r \in R} z_r \geq nv \right\} \tag{4}$$
The algorithm to solve (4) is the same developed to solve (3), properly adjusted to account for the additional constraint on the number of vehicles used. We describe the labeling algorithm in Sect. 4.1. To improve lower bound lb, we also add a relaxed form of the well-known subset-row inequalities [10] tailored to the PDPTWL so that they are *robust* in our pricing problem—we provide the details in Sect. 4.2.

4.1 Pricing Problem

As previously mentioned, the pricing problem is solved with a labeling algorithm combining the t-route state space relaxation with partial elementarity. Let $B \subseteq P$ be the set of customers selected to be visited at most once in each route (according to the partial elementarity principles). The labels of the labeling algorithm (corresponding to directed paths in G starting from the depot and consisting of feasible sequences of linking arcs and fragments) are defined as $\mathcal{S} = \{(S, t, i) \mid S \subseteq B, t \in [e_i, l_i], i \in V\}$, where the resources are $S \subseteq B$ (the set of customers of the set B visited), $t \in [e_i, l_i]$ (the arrival time at customer $i \in N$), and $i \in N$ (the last visited customer). We denote with $rc(S, t, i)$ the reduced cost of label $(S, t, i) \in \mathcal{S}$ w.r.t. dual variables $\pi \in \mathbb{R}^n$ associated with constraints (2b), and dual variable $\pi_0 \in \mathbb{R}_+$ associated with the constraint on the imposed number of vehicles. These reduced costs are computed according to the following resource extension functions:

If $i = 0$: label $(\emptyset, e_0, 0)$ is extended with each linking arc $(0, i) \in E^{0P}$ to generate label $(\emptyset, e_i, i)$ of reduced cost $rc(\emptyset, e_i, i) = C + c_{0i} - \pi_0$;

If $i \in P$: label (S, t, i) is extended with each fragment $f \in \Phi_i$ such that $t \leq l_f$ and $j \notin S$ for each $j \in P$ such that $a_{fj} = 1$, to generate label $(U, \max\{t, e_f\} + t_f, i + n)$, where $U = \{k \in B \mid k \in S \text{ or } a_{fk} = 1\}$; the reduced cost of the generated label is $rc(S, t, i) + c_f - \sum_{k \in P: a_{fk}=1} \pi_k$;

If $i \in D$: label (S, t, i) is extended with each linking arc $(i, j) \in E^{DP} \cup E^{D0}$ such that $j \notin S$ and $t + t_{ij} \leq l_j$ to generate label $(S, \max\{t + t_{ij}, e_j\}, j)$ of reduced cost $rc(S, t, i) + c_{ij}$.

To discard dominated labels, we apply the following dominance rule. Given two labels $(S^1, t^1, i^1), (S^2, t^2, i^2) \in \mathcal{S}$ such that $i^1 = i^2$, if $S^1 \subseteq S^2$, $t^1 \leq t^2$, and $rc(S^1, t^1, i^1) \leq rc(S^2, t^2, i^2)$, then label (S^2, t^2, i^2) is dominated and can be discarded. For the sake of space, we omit the details on the adjustments to the labeling algorithm to take valid inequalities described in Sect. 4.2 into account.

As to the definition of the set B of customers on which elementary is imposed, the set is empty in Step 2 described in Sect. 4. In Step 3, whenever no negative-reduced-cost route is found and no valid inequality is added, one of the customers visited more than once in the routes in solution in the restricted master (if such a customer exists) is added to the set B.

4.2 Valid Inequalities

One of the most popular sets of valid inequalities used to strengthen the linear relaxation of set partitioning formulations in the literature about routing problems is the subset-row inequalities introduced in [10]. These inequalities are a class of cuts, but they are often used in their simplest form, called *3-subset-row inequalities* (3SR). Applied to formulation (2) they read as:

$$[3SR] \qquad \sum_{\substack{r \in R: \\ a_{ri}+a_{rj}+a_{rk} \geq 2}} z_r \leq 1 \qquad \forall \{i, j, k\} \in P \tag{5}$$

The addition of the 3SR inequalities to the linear relaxation of (2) has undesirable consequences on the pricing problem: to take the dual variables of (5) into account, additional resources are necessary, hence the pricing problem becomes even more challenging. For this reason, 3SR inequalities are *non-robust cuts*.

In our bounding procedure, we separate a weaker version of 3SR cuts (that we call *Weak-3SR*) that are robust, thus maintaining the correctness and complexity of the solution method described in Sect. 4.1 to generate negative reduced-cost columns. In particular, we separate the following Weak-3SR cuts:

$$[\text{Weak-3SR}] \qquad \sum_{r \in R(i,j,k)} z_r \leq 1 \qquad \forall \{i, j, k\} \in P \tag{6}$$

where $R(i, j, k) \subseteq R$ is the set of routes that traverse one of the fragments that visit at least two of the three pickups i, j, k. The separation of cuts (6) can easily be performed by inspection.

4.3 *Master Problem Stabilization*

To speed up the column generation procedure by (hopefully) decreasing the number of calls to the pricing algorithm, we apply *in-out*, a well-known stabilization technique introduced in [2], to which we refer the reader for all the details. The idea of in-out is not to immediately generate columns of negative reduced-cost w.r.t. the current dual solution of the master problem $\boldsymbol{\pi} \in \mathbb{R}^n$, but to search for negative reduced-cost columns w.r.t. a vector $\boldsymbol{\pi}' \in \mathbb{R}^n$ that is in between $\boldsymbol{\pi}$ and an internal point of the dual space $\boldsymbol{\pi}^0 \in \mathbb{R}^n$; the search of negative reduced-cost columns w.r.t. $\boldsymbol{\pi}$ is performed only if no negative reduced-cost columns w.r.t. $\boldsymbol{\pi}'$ can be found. Both vectors $\boldsymbol{\pi}^0$ and $\boldsymbol{\pi}'$ can be modified during the column generation procedure.

5 Computational Results

We have tested our bounding procedure (hereafter referred to as BP) on the set of benchmark instances used in both [1, 3]—the latter is the state-of-the-art exact method for the PDPTWL. These instances were obtained by modifying the ones in [16]. The set of instances consists of 60 instances divided into six groups (called AA, BB, CC, DD, AA*, BB*) of 10 instances each; each group contains one instance for each $n \in \{30, 35, 40, 45, 50, 55, 60, 65, 70, 75\}$. In all instances, the vehicle's fixed cost C is 10,000.

We have coded BP in C language, using the Gurobi Optimizer (version 10.0.2) to solve the master problem, and have tested it on a machine Intel(R) Xeon(R) CPU E3-1220 V2 @ 3.10GHz, 16GB of RAM running Fedora Linux 37. After conducting some parameter tuning, we have decided to use the following parameter setting: (i) the maximum number of pickups subject to partial elementarity is 5 (i.e., $|B| = 5$); (ii) at each iteration of column generation in Steps 2 and 3 of BP, we add at most 20 columns of negative reduced-costs and, in Step 3, all violated Weak-3SR cuts; (iii) in the stabilization of the master problem, we use at most three dual vectors $\boldsymbol{\pi}'$ in the pricing algorithm before using $\boldsymbol{\pi}$.

Table 1 summarizes the features of the benchmark instances and compares the results achieved by BP and the exact method presented in [1] (referred to as AFB) on each of the six groups. Under Instances, column # indicates the number of instances in the group, column Q the vehicle capacity, column W the width of all time windows, and Δ the delay between the time window of each pickup i and its delivery $j = i + n$ (i.e., $e_j - e_i = l_j - l_i = t_{ij} + \Delta$). Under AFB, column opt indicates the number of instances solved to optimality by AFB with a time limit of one hour, $\mathsf{T}_{\mathsf{opt}}$ the average computing time in seconds (considering the time limit for the open instances), and gap the average percentage gap at the root node (only considering the instances solved to optimality). Under BP, column gap reports the average percentage gap at the root node (only considering the instances for which an optimal solution is known), column T_{lb} the average computing time in seconds, column elem the number of instances where all routes in the optimal solution corresponding to lb are elementary, and opt the number of instances where lb is the optimal value.

Table 1 shows that the average gaps achieved with BP are significantly lower than the gaps of AFB; this shows the tightness of the linear relaxation of (2) strengthened with Weak-3SRs (6). The computing times of BP are a few seconds, so an exact method based on BP is expected to be competitive with AFB. Such limited computing times are owed to both the effective pricing algorithm described in Sect. 4.1, which does not need to generate too many labels to price out columns of negative reduced cost, and the stabilization technique (see Sect. 4.3), which helps reduce the number of column-generation iterations. Finally, we can see that BP finds an optimal integer solution on 14 of the 60 instances, and all the routes in the final solution are elementary in 45 cases.

Table 1 Benchmark instances features, aggregated results, and comparison with the state-of-the-art exact method

Instances					AFB			BP			
Group	#	Q	W	Δ	opt	T_{opt}	gap	gap	T_{lb}	elem	opt
AA	10	22	60	45	10	1.3	0.014	0.001	2.0	10	6
BB	10	30	60	45	10	2.2	0.052	0.011	1.0	10	4
CC	10	18	120	15	7	1,265.6	0.060	0.021	6.6	3	1
DD	10	25	120	15	8	1,072.1	7.001	0.037	8.3	2	0
AA*	10	26	60	60	10	3.0	0.054	0.018	1.9	10	2
BB*	10	35	60	60	10	5.7	1.459	0.056	1.0	10	1
All	60				55					45	14

6 Conclusions

We have investigated the PDPTWL, which arises in distribution settings where loads are challenging to handle. We have described the problem using the concept of fragments and have developed a fragment-based MILP formulation. We have developed a new column generation algorithm to compute tight lower bounds to the problem. The pricing problem is solved with a labeling algorithm exploiting the t-route state space relaxation and partial elementarity, generating routes by alternating fragments and linking arcs. Dual stabilization techniques are applied to speed up the procedure, and a family of robust valid inequalities is separated to improve the bound. The final bounds are tight and can be achieved in a few seconds of computing time. Hence, our bounding procedure can be an essential component of an exact method for the PDPTWL. We will investigate how to close this final gap to achieve an optimal solution to the problem; to this end, we plan to combine column enumeration with branch-and-(cut-and)-price.

Acknowledgements We are grateful to the two anonymous referees for their careful reading and meaningful remarks that improved our paper. The work of the last two authors was supported by Ministero dell'Università e della Ricerca (MUR), PRIN 2022 project 20229ZWC97—"Revise and Enhance: Win-win INsights for home Delivery services" (REWIND).

References

1. Alyasiry, A.M., Forbes, M., Bulmer, M.: An exact algorithm for the pickup and delivery problem with time windows and last-in-first-out loading. Transp. Sci. **53**(6), 1695–1705 (2019)
2. Ben-Ameur, W., Neto, J.: Acceleration of cutting-plane and column generation algorithms: Applications to network design. Networks **49**(1), 3–17 (2007)
3. Cherkesly, M., Desaulniers, G., Laporte, G.: Branch-price-and-cut algorithms for the pickup and delivery problem with time windows and last-in-first-out loading. Transp. Sci. **49**(4), 752–766 (2015)
4. Cherkesly, M., Desaulniers, G., Laporte, G.: A population-based metaheuristic for the pickup and delivery problem with time windows and LIFO loading. Comput. Oper. Res. **62**, 23–35 (2015)
5. Christofides, N., Mingozzi, A., Toth, P.: State-space relaxation procedures for the computation of bounds to routing problems. Networks **11**(2), 145–164 (1981)
6. Desaulniers, G., Desrosiers, J., Solomon, M.M. (eds.): Column Generation. No. 978-0-387-25486-9 in Springer Books. Springer, September 2005
7. Desaulniers, G., Lessard, F., Hadjar, A.: Tabu search, partial elementarity, and generalized k-path inequalities for the vehicle routing problem with time windows. Transp. Sci. **42**(3), 387–404 (2008)
8. Dror, M.: Note on the complexity of the shortest path models for column generation in VRPTW. Oper. Res. **42**(5), 977–978 (1994)
9. Dumas, Y., Desrosiers, J., Soumis, F.: The pickup and delivery problem with time windows. Eur. J. Oper. Res. **54**(1), 7–22 (1991)
10. Jepsen, M., Petersen, B., Spoorendonk, S., Pisinger, D.: Subset-row inequalities applied to the vehicle-routing problem with time windows. Oper. Res. **56**(2), 497–511 (2008)

11. Liu, F., Gui, M., Yi, C., Lan, Y.: A fast decomposition and reconstruction framework for the pickup and delivery problem with time windows and LIFO loading. IEEE Access **7**, 71813–71826 (2019)
12. Parragh, S.N., Doerner, K.F., Hartl, R.F.: A survey on pickup and delivery problems. J. für Betriebswirtschaft **58**(1), 21–51 (2008)
13. Parragh, S.N., Doerner, K.F., Hartl, R.F.: A survey on pickup and delivery problems. J. für Betriebswirtschaft **58**(2), 81–117 (2008)
14. Rist, Y., Forbes, M.: A column generation and combinatorial benders decomposition algorithm for the selective dial-a-ride-problem. Comput. Oper. Res. **140**, 105649 (2022)
15. Rist, Y., Forbes, M.A.: A new formulation for the dial-a-ride problem. Transp. Sci. **55**(5), 1113–1135 (2021)
16. Ropke, S., Cordeau, J.-F.: Branch and cut and price for the pickup and delivery problem with time windows. Transp. Sci. **43**(3), 267–286 (2009)

Meal Delivery with Unmanned Aerial Vehicle: A Case Study from the University of Calabria

Alessia Ciacco, Martina Luzzi, and Edoardo Scalzo

Abstract In recent times, the popularity of food delivery services has hugely increased. This study delves into the prospect of employing Unmanned Aerial Vehicles for food delivery. In particular, the context of the application under examination is the campus area of the University of Calabria, where the canteens play a crucial role as food providers for students and employees. A bi-objective mixed integer linear programming model is formulated, with the aim of finding the optimal routes to serve customers. The goal is to minimize travel distances and travel time, enhancing overall process efficiency. The ε-constraint approach is used to solve the problem. The preliminary results obtained are promising and indicate the potential effectiveness of using drones for food delivery, within the university campus setting.

Keywords UAV · Meal delivery · Bi-objective optimization · Routing · ε-constraint method

1 Introduction

Meal delivery services have become popular with the rise of online platforms and mobile apps, that connect consumers with local restaurants and food providers. In addition, social distancing and the restrictions imposed by lockdown, due to the pandemic, have further contributed to the growth of this sector. Nowadays, food delivery is a global market worth more than 150 billion, having more than tripled since 2017 [1]. Basically, food delivery refers to the service of bringing prepared meals or food

A. Ciacco (✉) · M. Luzzi · E. Scalzo
Department of Mechanical, Energy and Management Engineering, University of Calabria, Rende, Italy
e-mail: alessia.ciacco@unical.it

M. Luzzi
e-mail: martina.luzzi@unical.it

E. Scalzo
e-mail: edoardo.scalzo@unical.it

M. Barbato et al. (eds.), *Beyond Frontiers of Operations Research: Emerging Technologies and Innovative Optimization Paradigms*, AIRO Springer Series 14,
https://doi.org/10.1007/978-3-031-85894-9_7

items from a restaurant or food establishment to the locations of customers, who can enjoy a wide variety of food options without leaving their homes or work offices. Additionally, restaurants and, more generally, food providers can expand their customer base and increase sales. Typically, the process of food delivery is composed of five steps: order placement, payment, food preparation, delivery, and receipt of the order. Initially, customers use a website or mobile app to browse menus, select items, and place an order from a restaurant or food provider. Subsequently, payment is usually made online through the platform using various methods, including credit/debit cards, digital wallets, or other payment options. Once the order is received, the restaurant or food establishment prepares the food according to the requests of the customer. Following this, a designated delivery person, often employed by the food delivery platform or the restaurant itself, picks up the prepared order and transports it to the specified delivery address indicated by the customer during the order placement. Lastly, the customer receives the food at his doorstep or another designated location. Considering the high volumes of this market and the huge number of people and operations involved in this process, it is important to manage and optimize every single activity. Indeed, the delivery phase represents a subject that has gained interest among scholars and practitioners. In the work proposed by [8], some optimization-based algorithms to solve the courier assignment (dynamic vehicle routing) and capacity management (offline shift scheduling) problems encountered in meal delivery operations are developed. In [10], a stochastic dynamic pickup and delivery problem, in which a fleet of drivers delivers food from a set of restaurants to serve customers, avoiding delays, is presented. The authors of [12] have provided an optimization algorithm for scheduling riders involved in meal delivery. In recent times, the advent of new green technologies, such as Unmanned Aerial Vehicles (UAVs), hereafter denoted also as drones, able to transport products without the direct presence of a human driver, has encouraged several researchers to investigate the possibility of employing drones to perform parcel deliveries in the logistics sector (e.g., [5, 7, 11]). Following these assumptions, this work addresses the use of UAVs for meal delivery, specifically within the context of the University of Calabria. This institution is recognized as one of the largest campuses in Italy, providing a distinctive environment for the implementation and study of such advanced logistical solutions. To the best of our knowledge, the use of drones to deliver meals has not yet been implemented in the real market, probably due to technology and regulatory constraints. However, in the last few years, some researchers have started to study the problem of delivering meals with UAVs. In [6], the strategic location of the charging stations for drones employed in the delivery of meals is discussed. In particular, an optimization model addressing the trade-off between the number of stations and the coverage of the demand is presented. A temporally discrete and spatially continuous mixed integer programming formulation for on-demand meal delivery using drones is proposed in [4]. The recent work [3] presents a mixed closed-open multi-depot routing and scheduling problem for homemade meal delivery incorporating drones and a crowd-sourced fleet. The authors developed a self-adaptive hyper-heuristic approach, based on a genetic algorithm, and a modified particle swarm optimization algorithm to solve the problem.

The aim of this work is to investigate the drone aided meal delivery problem in the context of a university campus. The key difference between the proposed research and the works that already exist in the literature is, basically, the application context. Indeed, the campus represents an area in which canteens and potential customers are precisely located. Since the campus area is limited and safe from air traffic, it is possible to use drones concretely. Currently, the sole available option for meals is to walk to one of the canteens situated within the campus area. Canteens frequently experience overcrowding, leaving limited time for meals. Therefore, it would be convenient to receive the meal directly in the office or in the green rest areas of the campus. Thus, a model formulation aimed at finding the optimal route for drones that deliver meals on campus, taking into account the congestion levels of the canteens during the day and the traveled distance to reach customers, is developed.

The remainder of the paper is structured as follows: Sect. 2 presents the formulation of the problem. In Sect. 3, the computational tests and the results analysis are made. Concluding remarks and insights for potential future research are condensed in Sect. 4.

2 Problem Statement

2.1 Characteristics of the Problem

In the considered scenario, the objective is to determine the optimal route for a drone to deliver meals from university canteens to customers located within the university campus, knowing that each canteen can be visited at most once by the drone during its route. It is assumed that the drone starts its route fully charged, with enough battery capacity to visit all the campus canteens once. A set of customers, denoted by N, distributed geographically within the urban area, and a set of canteens, V, are considered. This system is characterized by an origin depot 0 and a dummy node $n+1$ representing a copy of the depot 0 in which the route ends. The set $U := V \cup N \cup \{0, n+1\}$ encompasses all nodes. The workday of the canteens is divided into three time slots, represented by the set H, allowing, differentiation at various moments of the day based on the intensity of demands. The **first time slot** occurs from 12:00 to 14:00, coinciding with lunchtime. The **second time slot** spans from 14:00 to 19:00. The **third time slot**, starting around 19:00, extends until the close of daily operations, typically around 22:00, coinciding with dinner time. For each $h \in H$, these intervals are denoted by $[l^h, u^h]$. So, l^h is the earliest possible time to perform services during the specified time slot h, and u^h is the latest time at which meals must be placed onto the drone within the defined time slot h. Each customer has a designated time window within which the service must be provided, defined as the time interval $[a_i, b_i]$. The delivery service for customer i cannot be performed if the drone will arrive at i after instant b_i. However, if the drone arrives earlier than a_i, it waits until a_i before initiating the service. The service time associated with each customer i is s_i. Furthermore, w_i^h represents the service time associated with canteen $i \in V$ during the specific time slot $h \in H$. This parameter indicates the estimated amount of time it will take to prepare the food and load it into the drone at the canteen i during the particular time slot h. Let g_i be the quantity of meals required by customer $i \in N$. It is explicitly set to 0 for any node other than customers. It is assumed that the drone has a maximum capacity Q and travels at a constant speed v. The distance between two nodes i and j belonging to set U is denoted as d_{ij}, while the corresponding travel time is defined as $t_{ij} := d_{ij}/v$. Note that the distances d_{ij} satisfy the triangle inequality. This property guarantees a consistent and feasible spatial relationship between the nodes of the set U.

The variables used to define the model are:

- x_{ij}^h binary variable equal to 1, if the drone travels from node $i \in U$ to $j \in U$ picked up during time slot $h \in H$ and 0, otherwise;
- τ_i time instant at which the drone can depart from node $i \in U$;
- q_j drone load on arrival at the node $j \in U$;
- z_i quantity of meals that the drone takes from the canteen $i \in V$.

2.2 Mathematical Formulation

The problem is mathematically formulated as follows.

$$\min \sum_{i \in N} \tau_i \tag{1}$$

$$\min \sum_{h \in H} \sum_{i \in U} \sum_{j \in U} d_{ij} x_{ij}^h \tag{2}$$

s.t.

$$\sum_{h \in H} \sum_{j \in V} x_{0j}^h - \sum_{h \in H} \sum_{i \in N} x_{i,n+1}^h = 0 \tag{3}$$

$$\sum_{h \in H} \sum_{i \in V} x_{0i}^h = 1 \tag{4}$$

$$\sum_{h \in H} \sum_{j \in N} x_{ij}^h - \sum_{h \in H} \sum_{j \in N \cup \{0\}} x_{ji}^h = 0 \qquad \forall i \in V \tag{5}$$

$$\sum_{j \in U \setminus \{0\}} x_{ij}^h - \sum_{j \in U \setminus \{0, n+1\}} x_{ji}^h = 0 \qquad \forall i \in N, \forall h \in H \tag{6}$$

$$\sum_{h \in H} \sum_{j \in U \setminus \{0\}} x_{ij}^h = 1 \qquad \forall i \in N \tag{7}$$

$$\tau_i \geq (a_i + s_i) \sum_{h \in H} \sum_{j \in U \setminus \{0, n+1\}} x_{ji}^h \qquad \forall i \in N \tag{8}$$

$$\tau_i \leq (b_i + s_i) \sum_{h \in H} \sum_{j \in U \setminus \{0, n+1\}} x_{ji}^h \qquad \forall i \in N \tag{9}$$

$$\tau_j \geq \tau_i + t_{ij} + s_j + M(x_{ij}^h - 1) \qquad \forall i \in U, \forall j \in N \cup \{n+1\}, \forall h \in H \tag{10}$$

$$\tau_j \geq \tau_i + t_{ij} + w_j^h + M(x_{ij}^h - 1) \qquad \forall i \in U, \forall j \in V, \forall h \in H \tag{11}$$

$$\tau_j \geq l^h + w_j^h + M(x_{ij}^h - 1) \qquad \forall i \in N \cup \{0\}, \forall j \in V, \forall h \in H \tag{12}$$

$$\tau_j \leq u^h + w_j^h + M(1 - x_{ij}^h) \qquad \forall i \in N, \forall j \in V, \forall h \in H \tag{13}$$

$$q_i = 0 \qquad i = 0, n+1 \tag{14}$$

$$q_j \leq q_i + z_i + M(1 - x_{ij}^h) \qquad \forall i \in V, \forall j \in N, \forall h \in H \tag{15}$$

$$q_j \leq q_i - g_i + M(1 - x_{ij}^h) \qquad \forall i \in N, \forall j \in U \setminus \{0\}, \forall h \in H \tag{16}$$

$$z_i \leq Q - q_i \qquad \forall i \in V \tag{17}$$

$$\sum_{i \in V} z_i = \sum_{j \in N} g_j \tag{18}$$

$$x_{ij}^h \in \{0, 1\} \qquad \forall i \in U, \forall j \in U, \forall h \in H \tag{19}$$

$$\tau_i \geq 0 \qquad \forall i \in U \tag{20}$$

$$q_i \in [0, Q] \cap \mathbb{Z} \qquad \forall i \in U \tag{21}$$

$$z_i \in [0, Q] \cap \mathbb{Z} \qquad \forall i \in V. \tag{22}$$

The expression (1) minimizes the time dedicated to meal delivery aims at optimizing the efficiency of the process, focusing on mitigating delays and streamlining the temporal aspects of the delivery system. The expression (2), known as routing function, aims to minimize the total distance incurred by the routes charted by the drone. It optimizes the spatial trajectories, seeking to find the most convenient paths for meal delivery across the university campus. Thus, (2) depends on the distance travelled by the drone. Constraint (3) ensures that the drone departs from the depot node and arrives at the destination node $n + 1$. Constraint (4) guarantees that the drone departs exactly once from the depot. Equations (5) and (6) are the drone flow conservation constraints. Constraints (7) mandates that all customers must be served once and only once. Constraints (8) and (9) ensure compliance with the time windows $[a_i, b_i]$. Constraints (10) guarantees that if the route from $i \in U$ to $j \in N \cup \{n+1\}$ is performed by the drone, the instant at which the drone can depart from customer j is not earlier than the moment when the drone departs from node i, considering the travel time of the arc and the service time of customer j. Constraints (11) states that, if arc (i, j) is travelled by the drone picking up from canteen $j \in V$ in slot $h \in H$, the time in which the drone arrives at canteen j must be not less than the time in which the drone arrives at node $i \in U$, added to the travel time from node i to node j and the service time of canteen w_j^h. Constraints (12) and (13) express that for each canteen, the service begins within the temporal interval between l^h and u^h plus the service time of each canteen w_j^h and must be concluded before u^h. Constraints (14) express that the drone load at the depot must be 0. Constraints (15) guarantee that, if the arc from canteen i to customer j is travelled by the drone in slot h, its load before arriving at node j must not be greater than the drone load at node i plus the amount taken at canteen i. Constraints (16) ensure that if the arc from customer i to node j is travelled by the drone in slot h, its load before arriving at node j must not be greater than the drone load at customer i minus the quantity taken from customer i. Constraints (17) ensures that the drone at the canteens cannot take more meals than its current capacity. Constraint (18) expresses that the total amount of meals picked up by the drone at the canteens must be exactly equal to the total demand of the customers. Constraints (19)–(22) define the domain of the decision variables. The objective function (1) and customers time window constraints (8) and (9) are integrated to balance operational efficiency together with the temporal restrictions of customers for the meal delivery. Indeed, the objective function (1) aims to minimize arrival time at the customers, optimizing the delivery process for greater overall efficiency. Simultaneously, the implementation of customer time window constraints is crucial for customer satisfaction, as ignoring them could cause inconvenience by compromising the customer experience. In other words, the conjunction implementation of (1), (8), and (9) allows to serve customers not only within their time windows but also as soon as possible, avoiding delays and enhancing overall efficiency and service level.

2.3 *Solution Approach*

The solution approach adopted to solve the proposed problem is ε-constraint [2]. Basically, it consists in solving a multi-objective programming problem by considering one objective function at time. In this case, this procedure comprises the following phases. **First phase**: solving the problem P1 composed by the function (1) and constraints from (3) to (22). **Second phase**: solving the problem P2 composed by the function (2) and constraints from (3) to (22), incorporating the additional constraint ε-constraint defined as: $\sum_{i \in N} \tau_i \leq (\varepsilon + 1)\ f^*$, where f^* represents the optimal solution obtained solving P1, while ε is a parameter set to determine maximum acceptable deterioration for function (1).

3 Computational Experiments

The mathematical model was coded in C++ and solved with CPLEX 12.10, imposing a time limit of 3600 s (s) for each instance and each problem, allowing CPLEX to be used in a heuristic manner if finding an optimal solution requires more time. All computational tests were conducted using an Apple M2 Max processor with CPU 12-core and GPU 38-core and 96 GB LPDDR5 of RAM running macOS Ventura 13.3.

3.1 Instances Generation

To validate the proposed model several tests were conducted considering two groups of instances. The first set of instances is obtained from the well-known Solomon benchmarks [9] for vehicle routing problems with time windows. A depot 0 and its copy $n + 1$ are set equal to the original depot of the Solomon instance. Customers and canteens are selected randomly from the nodes of the instance, and their coordinates remain unchanged. The time window and the service time of customers remain the same as the original ones. The various time durations, expressed in minutes and discussed in the paper, have all been converted into Solomon time units. Thus, given that the real planning horizon considered spans 10h (from 12:00 to 22:00), 1 real-world minute is equivalent to $T_{max}/600$ in Solomon time, where T_{max} is the Solomon planning horizon. Three service times w_i^h are generated for each canteen, estimating randomly among the intervals $[10, 20]$ for $h = 1$, $[2, 5]$ for $h = 2$, and $[10, 15]$ for $h = 3$. These intervals are calculated on the basis of information gathered from a survey submitted to people who regularly use the canteen service at the University of Calabria. The extremes of the intervals are expressed in minutes and give an idea of the congestion level of each time slot: $h = 1$ for high congestion, $h = 2$ for low congestion, and $h = 3$ for medium congestion. The quantity of meals g_i, $i = 1, \ldots, |C|$, requested by each customer is randomly determined within the range of 1 and $max(1, min(Q, |V| \times Q - |N| + 1 - \sum_{h=1,\ldots,i} g_h))$. This ensures the overall feasibility of the instance. For all other nodes in the network, such as the depot and canteens, the meal quantity is consistently set to zero. The capacity of the drone Q is consistently maintained at 5. For simplicity, values of c and v are set equal to 1. The drone distance d_{ij} is calculated according to the Euclidean metric.

The second set of instances is derived by considering the geographic layout of the University of Calabria. As can be seen from Fig. 1, the campus features five canteens, depicted by yellow circles. The position of the depot, highlighted in the figure with the red cross, is chosen near the electrical generators of the campus. Within the university, customers, including students, researchers, and technical administrative staff, are distributed across diverse locations such as offices, laboratories, green areas, and residences, as represented in Fig. 1 by red and green polygons. The real coordinates of 20 possible customers disseminated on campus, along with the locations of 5 canteens and a depot, have been gathered. The planning horizon T_{max} is 1236, and to maintain the Solomon proportion, the service time s_i is equal to 10. Customer time windows are obtained by considering a duration between 30 and 55 min. The values of w_i^h, g_i, Q, c, v, l^h, u^h and d_{ij} are set as previously discussed.

3.2 Numerical Results

A total of 40 testing instances are considered, 20 from the Solomon benchmark and 20 from the Real class. For each instance of Solomon, three canteens are

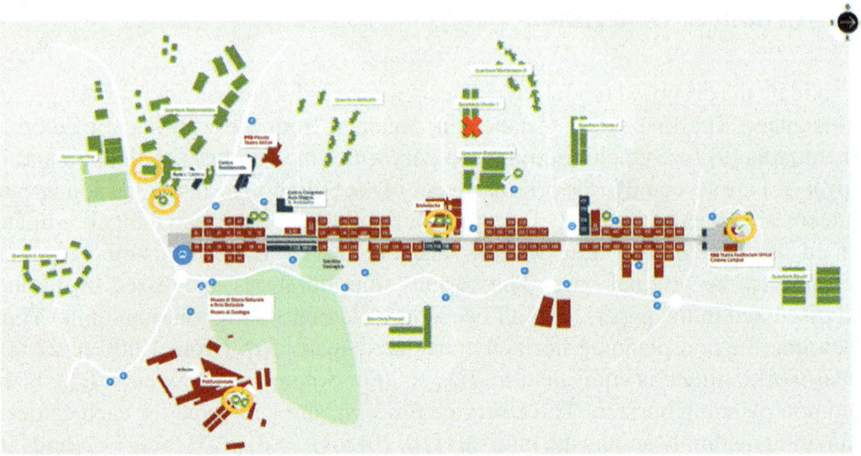

Fig. 1 Map of the University of Calabria. *Source* https://www.unical.it/campus/visita-il-campus/mappa/

considered. Each set includes a subset of 10 instances with 5 customers, indicated as C5, and a subset of 10 instances with 10 customers, indicated C10. The results of the computation are presented in Tables 1 and 2, which reports the average, over the number of instances considered, of the execution times for P1 and P2, the value of the objective function of P1 and P2, denoted by Ob_1 and Ob_2, respectively, and the relative CPLEX gap of P1 and P2, denoted by Gap_{P1} and Gap_{P2}, varying the value of ε. As can be seen from the results, as the ε increases, there is a proportional worsening of the Ob_1 and an improvement of the Ob_2, as expected. Regarding the gap, CPLEX proves to be highly reliable in seeking the optimal solution for problem P2 in all instances, both Solomon and real. For P1, CPLEX consistently reaches the optimal solution in instances with 5 customers. In Solomon instances with 10 customers, CPLEX identifies a feasible sub-optimal solution only in two cases, with an average relative class gap of 14.35%, specifically the punctual values of the two gaps are 85.80 and 57.69% respectively. Whereas, in the remaining 8 instances, it found an optimal solution. Additionally, in real instances, CPLEX obtains the optimal solution in all situations except one, with an average relative gap of 0.20%.
The graph reported in Fig. 2 shows that in Solomon instances, the improvement percentage of the routing function compared to that obtained with $\varepsilon = 0$, named as *imp pct*, is observable up to $\varepsilon = 0.5$, while in real instances, with 5 and 10 customers, an increase of 3% is recorded only for $\varepsilon = 0.1$. This aspect emphasizes the critical importance of the accurate selection of ε to optimize the performance of the system, ensuring an optimal balance between the objectives of the problem. Regarding execution time, similar trends are observed across both classes. Most of the time, which amounts to an average of 91%, is spent solving problem P1 in Solomon instances, both with 5 and 10 customers. While, in real instances, the percentage of execution time spent on problem P1 decreases to about 70% for instances with 5 customers

Table 1 Average results obtained on the instances belonging to the *Solomon* class

	ε	0.0	0.1	0.3	0.5	0.7	0.9	1.0
C5	$Time_{P1}(s)$	2.42	2.42	2.47	2.44	2.41	2.41	2.41
	$Time_{P2}(s)$	0.42	0.36	0.32	0.17	0.15	0.18	0.19
	Ob_1	1631.62	1760.11	1970.14	2015.63	2037.90	2098.78	2083.02
	Gap_{P1}	0.00%	0.00%	0.00%	0.00%	0.00%	0.00%	0.00%
	Ob_2	223.80	210.20	205.00	199.90	199.90	199.90	199.90
	Gap_{P2}	0.00%	0.00%	0.00%	0.00%	0.00%	0.00%	0.00%
C10	$Time_{P1}(s)$	884.59	883.14	883.54	883.47	883.52	883.42	883.40
	$Time_{P2}(s)$	303.62	61.42	57.21	50.85	50.50	46.76	50.69
	Ob_1	6147.2	6761.92	7614.41	7623.82	7596.2	7427.7	7539.3
	Gap_{P1}	14.35%	14.35%	14.35%	14.35%	14.35%	14.35%	14.35%
	Ob_2	476.9	425	419.3	419.2	417.9	417.9	417.9
	Gap_{P2}	0.00%	0.00%	0.00%	0.00%	0.00%	0.00%	0.00%

Table 2 Average results obtained on the instances belonging to the *Real* class

	ε	0.0	0.1	0.3	0.5	0.7	0.9	1.0
C5	$Time_{P1}(s)$	1.14	1.13	1.13	1.13	1.13	1.13	1.13
	$Time_{P2}(s)$	0.63	0.48	0.44	0.48	0.44	0.53	0.47
	Ob_1	3065.70	3286.82	3215.86	3239.10	3252.16	3229.90	3255.48
	$RelGap_{P1}$	0.00%	0.00%	0.00%	0.00%	0.00%	0.00%	0.00%
	Ob_2	392.30	383.10	383.10	383.10	383.10	383.10	383.10
	$RelGap_{P2}$	0.00%	0.00%	0.00%	0.00%	0.00%	0.00%	0.00%
C10	$Time_{P1}(s)$	691.51	692.03	691.08	690.85	691.08	690.89	690.42
	$Time_{P2}(s)$	142.62	104.90	76.42	81.52	94.59	87.95	106.62
	Ob_1	6426.70	6702.80	6660.96	6665.24	6708.64	6691.44	6679.30
	$RelGap_{P1}$	0.20%	0.20%	0.20%	0.20%	0.20%	0.20%	0.20%
	Ob_2	782.80	758.40	758.40	758.40	758.40	758.40	758.40
	$RelGap_{P2}$	0.00%	0.00%	0.00%	0.00%	0.00%	0.00%	0.00%

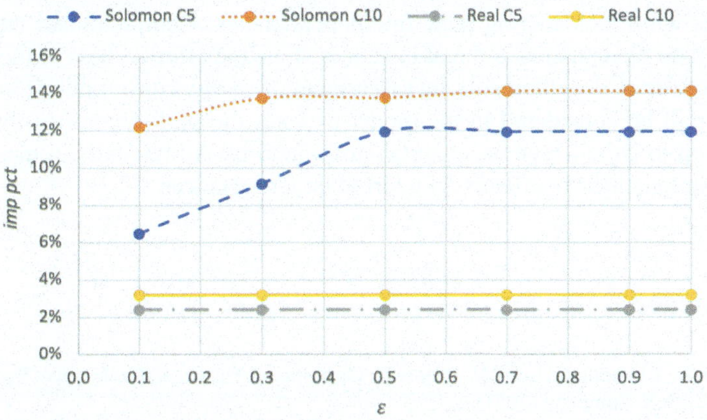

Fig. 2 Percentage improvement of the Ob_2 function compared to that obtained with $\varepsilon = 0$

and to 88% for those with 10 customers. Furthermore, in both classes with either 5 or 10 customers, as the *imp pct* increases, the percentage of execution time spent to solve the problem P2 respect to the time required to solve both P1 and P2 decreases. Moreover, from the Pareto frontiers, calculated considering the average values of the obtained solutions, it can be noticed that for $\varepsilon \in \{0.1, 0.3, 0.5\}$ there are the greatest number of Pareto-optimal solutions. Additional experiments were conducted on 40 instances, divided between the Solomon and Real classes, each with 15 customers. These instances were generated as previously explained in Sect. 3.1. However, the related results are not included in the tables because a feasible solution is not found within the 2-hour time limit imposed for each problem. This suggests the necessity of implementing a more effective approach to solve larger instances. It is notable that the values of the routing distance remain unaltered, varying the value of ε for real instances. This is since the customer time windows, generated based on the answers provided by the survey, have few overlaps among them. Consequently, the space of feasible solutions is relatively limited, and regardless of the value of ε, the drone is likely constrained to follow the same path, or very similar paths, to meet the delivery requirements within the established time windows. This results in a lack of change in the value of Ob_2 for $\varepsilon > 0$.

4 Conclusions

In conclusion, this study addresses the challenge of optimizing meal delivery systems using UAVs. The proposed mathematical model integrates multiple objectives, balancing the efficiency of meal delivery processes and minimizing spatial trajectories identifying effective paths for meal delivery across a university campus. The ε-constraint approach allows flexibility in optimizing individual objectives while considering their interdependencies. Computational experiments conducted on benchmarks and real instances show the effectiveness of the proposed model. For future developments, an assessment of energy consumption during drone delivery will be considered in the model to asses the performance of the delivery system and to enhance environmental sustainability. Additionally, a metaheuristic will be implemented to effectively address larger instances.

References

1. Ahuja, K., Chandra, V., Lord, V., Peens, C.: Ordering in: The Rapid Evolution of Food Delivery. McKinsey & Company (2021)
2. Chankong, V., Haimes, Y.Y.: Multiobjective Decision Making: Theory and Methodology. Courier Dover Publications (2008)
3. Hamid, M., Nasiri, M.M., Rabbani, M.: A mixed closed-open multi-depot routing and scheduling problem for homemade meal delivery incorporating drone and crowd-sourced fleet: a self-adaptive hyper-heuristic approach. Eng. Appl. Artif. Intell. **120**, 105876 (2023)
4. Liu, Y.: An optimization-driven dynamic vehicle routing algorithm for on-demand meal delivery using drones. Comput. Oper. Res. **111**, 1–20 (2019)
5. Macrina, G., Pugliese, L.D.P., Guerriero, F., Laporte, G.: Drone-aided routing: a literature review. Transp. Res. Part C: Emerg. Technol. **120**, 102762 (2020)
6. Pinto, R., Zambetti, M., Lagorio, A., Pirola, F.: A network design model for a meal delivery service using drones. Int. J. Logist. Res. Appl. **23**(4), 354–374 (2020)
7. Pugliese, L.D.P., Guerriero, F., Macrina, G.: Using drones for parcels delivery process. Procedia Manufact. **42**, 488–497 (2020)
8. Reyes, D., Erera, A., Savelsbergh, M., Sahasrabudhe, S., O'Neil, R.: The meal delivery routing problem. Optim. Online **6571** (2018)
9. Solomon, M.M.: Algorithms for the vehicle routing and scheduling problems with time window constraints. Oper. Res. **35**(2), 254–265 (1987)
10. Ulmer, M.W., Thomas, B.W., Campbell, A.M., Woyak, N.: The restaurant meal delivery problem: dynamic pickup and delivery with deadlines and random ready times. Transp. Sci. **55**(1), 75–100 (2021)
11. Wang, Z., Sheu, J.-B.: Vehicle routing problem with drones. Transp. Res. Part B: Methodol. **122**, 350–364 (2019)
12. Xue, G., Wang, Z., Wang, G.: Optimization of rider scheduling for a food delivery service in o2o business. J. Adv. Transp. **1–15**, 2021 (2021)

An Effective Matheuristic Approach to Solve Nurse Rostering Problem

Cristian Belfiore

Abstract The Nurse Rostering Problem is a very hard staff scheduling problem that aims to assign a set of nurses to shifts within a scheduling horizon. Nurses are characterized by one or more skills, are hired through a contract that regulates their work, and can express preferences regarding days off or work shifts. A preferred coverage is defined for each shift in the planning horizon, which represents the number of nurses the hospital would like to have. We address here the version of the problem proposed in 2014 by Curtois through a revised version of FiNeMath matheuristic. Preliminary results showed that the heuristic can determine high-quality solutions in a very short time for medium and small instances.

Keywords Nurse Rostering Problem · Discrete optimization · Matheuristic · Timetabling

1 Introduction

Staff rostering is a major issue in all organizations. Within healthcare facilities, inappropriate rostering can cause a reduction in the quality of care, can negatively impact nurses' safety and their physical and psychological health, and can increase turnover, absenteeism, and overall costs [8]. The nurse rostering problem (NRP) aims to assign nurses to predetermined shifts within a fixed scheduling horizon. Assignments must satisfy a set of constraints derived from internal hospital rules and legal requirements and must consider the preferences expressed by nurses. The goal is, in general, to ensure adequate shift coverage while maximizing staff satisfaction and balance in workloads.

C. Belfiore (✉)
de-Health Lab—Laboratory of Decision Engineering for Health Care Services, Department of Mechanical, Energy and Management Engineering, University of Calabria, Rende, Italy
e-mail: cristian.belfiore@unical.it

M. Barbato et al. (eds.), *Beyond Frontiers of Operations Research: Emerging Technologies and Innovative Optimization Paradigms*, AIRO Springer Series 14,
https://doi.org/10.1007/978-3-031-85894-9_8

Computationally, the complexity of the NRP depends on the specific constraints considered in formulating the problem [15]. However, although there are instances solvable in polynomial time [15], real instances are very often classifiable among *NP-hard* problems. As a matter of fact, den Hartog et al. [7] showed that instances of NRP that consider preferred coverage for each shift, days off request, and forbidden sequences, are part of this category in the strong sense when there are only four types of shifts (three effective shifts and one rest shift). Therefore, it is critical to design solution approaches that can achieve good solutions quickly, both for practical reasons related to increasing effectiveness and efficiency in healthcare and for the boost that this can give to solving other problems that have a different nature but the same structure [3].

In this paper, we consider the version of the NRP proposed in [6] and develop a solution approach based on combining mathematical programming with heuristic techniques, already successfully applied to other problems [10]. This matheuristic yields promising results.

The remainder of the paper is organized as follows: in Sect. 2, a brief review of the most relevant literature on NRP formulations and solution approaches is reported. In Sect. 3, the specific problem addressed is described and a mathematical formulation is discussed. In Sect. 4, the solution approach used is analyzed. In Sect. 5, the results of the computational experiments conducted are presented, and finally, in Sect. 6, conclusions and some suggestions for possible future research are discussed.

2 Related Work

The NRP problem has been extensively studied by the scientific community over the years [1, 12, 19], and numerous formulations for that have been proposed as several solution approaches have been designed. Researchers have also taken a strong interest in the problem because of the international competitions that were held in 2010 and 2014.

The version of the problem proposed in the first International Nurse Rostering Competition (INRC-I) [11] considers nurses characterized by a single among a given set of skills and a single-stage (or "static") approach in which all information is assumed known with certainty at the beginning of the planning horizon.

The version of the NRP proposed for the second International Nurse Rostering Competition (INRC-II) considers, in contrast to its predecessor, a multi-stage (or "dynamic") version of the problem where the planning horizon is divided into week-long "stages" in each of which the problem is to be solved, and the overall evaluation of the solution is made only at the end of the entire planning [2]. This way of modeling the problem allows a better representation of what happens in reality, where decisions made previously influence decisions that must be made for the future. INRC-II was won by [18] who proposed a solution approach based on reformulating the problem into a multi-commodity flow MILP formulation based on state-expanded networks.

Over the years, additional datasets have been made available to the scientific community. One of them is known as the Shift Scheduling Benchmark Dataset [5], which is used in this paper and presented in Sect. 3. We chose to test our methodology on these instances because they accurately reflect real-world requirements and different possible scheduling scenarios [4]. Several researchers have worked on this dataset over time. A hybrid algorithm that combines integer programming with a variable neighborhood search approach was developed in [17]. The method has four phases. In the first phase, a greedy heuristic is employed to generate an initial feasible solution; in the second phase, an IP solver using a variable neighborhood descent algorithm attempts to improve the solution; in the third phase, a ruin and recreate framework is implemented to search for further improvements. Finally, an IP solver is used to refine the found solution. Another work is [15], where the authors empirically evaluated the impact of each set of constraints on the computational complexity of the problem through a series of experiments. On the basis of this evidence, they proposed a model based on a reformulation of the constraints related to the successive assignments in the problem of determining a feasible path in an oriented graph. In [16], a solution methodology derived from Branch and Price was proposed. The approach differs from it in two aspects: firstly, the column generation problem is not solved exactly but approximately, and secondly, a diving heuristic is used instead of fully exploring the tree. More recently, several authors worked on the same dataset. Among them, [3, 4] obtained relevant results using methods that combine traditional operations research methods with artificial intelligence techniques.

3 Problem Description

In the NRP there are a set of nurses with whom one or a set of skills may be associated. Nurses are hired through a contract that regulates their activities and can express preferences on days off or work shifts. The planning horizon is divided into days and each day into shifts. The healthcare facility defines a preferred coverage for each shift that represents the number of nurses that should be had to cover the demand without wasting resources. From hospital's internal rules and labor contracts, many different constraints that must be satisfied can be derived (for example, the maximum number of weekends a nurse can work in the planning horizon or the minimum and maximum number of consecutive working days). The goal is to assign nurses to shifts while respecting internal rules and contractual constraints by maximizing the nurses' preferences and minimizing the number and extent of understaffing and overstaffing situations.

The integer formulation for the NRP version here addressed as Model (1)–(12), that uses both hard constraints (which must necessarily be satisfied by a feasible solution) and soft constraints (which can be violated but with a negative impact on the quality of the solution), was proposed in [6]. Table 1 shows and explains the data structures, the parameters, and the variables necessary to formulate the problem.

Table 1 Set, parameters and variables used in model formulation

Notation	Definition
I	Set of nurses
h	Length in the planning horizon
$D = \{1, .., h\}$	Set of days in the planning horizon
$W = \{1, .., h/7\}$	Set of weekends in the planning horizon
T	Set of shift types
R_t	Set of shift types that cannot be assigned immediately after shift type t
N_i	Set of days that nurse i cannot work
l_t	Length in minutes of type shift t
m_{it}^{max}	Maximum number of shifts of type t that can be assigned to nurse i
b_i^{min}	Minimum number of minutes that nurse i must be assigned
b_i^{max}	Maximum number of minutes that nurse i can be assigned
c_i^{min}	Minimum number of consecutive shift that nurse i must work
c_i^{max}	Maximum number of consecutive shift that nurse i must work
o_i^{min}	Minimum number of consecutive days off that nurse i can be assigned
a_i^{max}	Maximum number of weekends that nurse i can work
q_{idt}	Penalty if shift type t is not assigned to employee i on day d
p_{idt}	Penalty if shift type t is assigned to nurse i on day d
u_{dt}	Preferred total number of nurses assigned shift type t on day d
v_{dt}^{min}	Weight if below the preferred cover for shift type t on day d
v_{dt}^{max}	Weight if exceeding the preferred cover for shift type t on day d
x_{idt}	1 if nurse i is assigned shift type t on day d, 0 otherwise
k_{iw}	1 if nurse i works on weekend w, 0 otherwise
y_{dt}	Total below the preferred cover for shift type t on day d
z_{dt}	Total above the preferred cover for shift type t on day d

$$\sum_{t \in T} x_{idt} \leq 1 \qquad \forall\, i \in I,\ d \in D \quad (1)$$

$$x_{idt} + x_{i(d+1)u} \leq 1 \qquad \forall\, i \in I,\ d \in \{1, .., (h-1)\},\ t \in T,\ u \in R_t \quad (2)$$

$$\sum_{d \in D} x_{idt} \leq m_{it}^{max} \qquad \forall\, i \in I,\ t \in T \quad (3)$$

$$b_i^{min} \leq \sum_{d \in D} \sum_{t \in T} l_t x_{idt} \leq b_i^{max} \qquad \forall\, i \in I \quad (4)$$

$$\sum_{j=d}^{d+c_i^{max}} \sum_{t \in T} x_{ijt} \leq c_i^{max} \qquad \forall\, i \in I,\ d \in \{1, .., (h - c_i^{max})\} \quad (5)$$

$$\sum_{t \in T} x_{idt} + \left(s - \sum_{j=d+1}^{d+s} \sum_{t \in T} x_{ijt} \right) + \sum_{t \in T} x_{i(d+s+1)t} > 0$$

$$\forall i \in I,\ s \in \{1, ..., (c_i^{min} - 1)\},\ d \in \{1, ..., (h - (s + 1))\} \quad (6)$$

$$\left(1 - \sum_{t \in T} x_{idt}\right) + \sum_{j=d+1}^{d+s} \sum_{t \in T} x_{ijt} + \left(1 - \sum_{t \in T} x_{i(d+s+1)t}\right) > 0$$

$$\forall i \in I,\ s \in \{1, ..., (o_i^{min} - 1)\},\ d \in \{1, ..., (h - (s + 1))\} \quad (7)$$

$$k_{iw} \leq \sum_{t \in T} x_{i(7w-1)t} + \sum_{t \in T} x_{i(7w)t} \leq 2k_{iw} \qquad \forall i \in I,\ w \in W \quad (8)$$

$$\sum_{w \in W} k_{iw} \leq a_i^{max} \qquad \forall i \in I \quad (9)$$

$$x_{idt} = 0 \qquad \forall i \in I,\ d \in N_i,\ t \in T \quad (10)$$

$$\sum_{i \in I} x_{idt} - z_{dt} + y_{dt} = u_{dt} \qquad \forall d \in D,\ t \in T \quad (11)$$

$$Minimize \ \sum_{i \in I} \sum_{d \in D} \sum_{t \in T} q_{idt} (1 - x_{idt}) + \sum_{i \in I} \sum_{d \in D} \sum_{t \in T} p_{idt} x_{idt} + \\ + \sum_{d \in D} \sum_{t \in T} y_{dt} v_{dt}^{min} + \sum_{d \in D} \sum_{t \in T} z_{dt} v_{dt}^{max} \quad (12)$$

Constraints (1) preclude assigning multiple shifts to a nurse on the same day. Constraints (2) exclude incompatible shift sequences from the feasible solutions (for example a nurse cannot work one day on the night shift and the next day on the morning shift). Constraints (3) set an upper bound on the number of shifts of each type to which each nurse can be assigned. Constraints (4) define a lower bound and an upper bound on the number of minutes that a nurse must work in the planning horizon. Constraints (5) impose a limit to the maximum number of consecutive shifts (and therefore days) that each nurse can work without days off. Constraints (6) and Constraints (7) impose for each nurse, respectively, a lower bound on the minimum number of consecutive days worked and on the minimum number of consecutive days off (see [6] for more information on how these constraints work). Constraints (8) and (9) impose the maximum number of weekends that a nurse can work (a weekend is considered worked if the nurse works on Saturday, Sunday, or both). Constraints (10) implement mandatory days off: each nurse is associated with a list of days in which he cannot work and in which, therefore, he must not be assigned to any shift. Constraint (11) quantifies, for a specific shift type t in day d, the number of nurses missing or in excess respect to the preferred cover.

The objective function (12) is made up of four terms: the first two allow maximizing the respect of the nurses' requests by attributing a penalty equal to q_{idt} for each non-working shift preference expressed by the nurses that are not respected and a penalty equal to p_{idt} for each work shift preference expressed by nurses not respected; note that different values of q_{idt} and p_{idt} can be defined for different nurses, different days in the planning horizon and different shift types. This is equivalent to assigning a priority to the requests received. The last two terms, however, consider the penalties due to overstaffing and understaffing situations. Specifically, for each day of the planning horizon and each shift, a penalty equal to v_{dt}^{min} for each nurse missing with respect to the preferred coverage demand and a penalty equal to v_{dt}^{max} for each nurse exceeding it is counted.

4 Solution Approach

The solution approach developed and implemented in this work is derived from the FiNeMath matheuristic, proposed in [10]. This solution approach simultaneously exploits a Fix and Optimize Heuristic (F&O), the Very Large Scale Neighborhood approach, and an integer linear programming solver allowing to obtain excellent quality solutions in a limited time by cooperating these three elements. The F&O heuristic was introduced in [9] for solving large instances of multiple type bus scheduling problems with multiple depots. This procedure is based on the decomposition of the original problem into several simpler subproblems that can be optimally solved. The solutions of those subproblems, properly combined, yield a solution to the original problem. Neighborhood Search Algorithms aim to achieve a good solution by generating iteratively a set of solutions close to the current solution among which one better is searched; when found, this becomes the new current solution and the procedure is repeated. Large neighborhood search (LNS) [13] is an iterative approach where the neighborhood is generated through the destruction of a portion of the current solution and, subsequently, a new feasible solution, hopefully better but still not worse than the starting solution, is identified through a heuristic. This methodology is effective and efficient only if the destruction method works well: if too much of the current solution is destroyed, the time required to reconstruct a feasible solution may be too high; conversely, if too little of the starting solution is destroyed, reconstruction is fast but is carried out in a too small search space where it is difficult to find a lower cost solution.

The contribution of this work is threefold. Firstly, we have used the framework provided by FiNeMath and derived from it a methodology useful for solving a NRP that, to the best of our knowledge, has never been addressed by anyone else in the literature. We have developed a new technique for determining a good initial feasible solution, based on the relaxation of some constraints that have a significant impact on the solution time of the problem. Finally, we have defined a destruction operator based on the complete destruction of assignments in groups of days, not necessarily consecutive, that is both efficient and effective.

The version of FiNeMath implemented here is based on two steps: the first step aims to identify a good initial feasible solution, and the second step aims to improve the initial solution by solving iteratively a series of smaller subproblems. The procedure is described in detail below.

4.1 Step 1: Find an Initial Feasible Solution

Let P the NRP formulation; let P^{Rel} the relaxed problem obtained from P by removing Constraints (6) and (7) related to the minimum number of consecutive days and days off; let T^1_{max} and T^2_{max}, be two time limits; let α_i be a set of integer value initially equal to $c_i^{max} \forall\, i \in I$.

Smet [14] showed that by removing Constraints (6) and (7) the computational time required to solve the problem proposed in [5] decreases significantly. Inspired by this consideration, the initial solution for P is determined by solving at optimum P^{Rel}. Assignments that in the optimal P^{Rel} solution satisfy the relaxed constraints (i.e. those that for a given nurse i extend for at least c_i^{min} consecutive days, followed by at least o_i^{min} consecutive days off) are identified for all nurses. These assignments were added as constraints in a new instance of the problem P, named P' (note that P' is a reduced problem of P). If the number of consecutive assignments for some nurses is greater than the upper bound α_i, only the last c_i^{min} consecutive assignments immediately preceding the days off are to be blocked for that. The initial feasible solution is obtained by solving the problem P' with an exact solver. If an optimal solution or at least a feasible solution for P' has not been found by T^1_{max}, or if P' has no solution, the procedure is repeated by decrementing to the non-negative integer immediately preceding the value of all α_i. If none of the at most $max_{i \in I}\{c_i^{max}\}$ iterations has produced a feasible solution, a starting feasible solution (that usually has a high cost), is obtained by solving through the solver a problem P'' that has the same constraints of P but considers a constant objective function. Figure 1 summarizes this procedure.

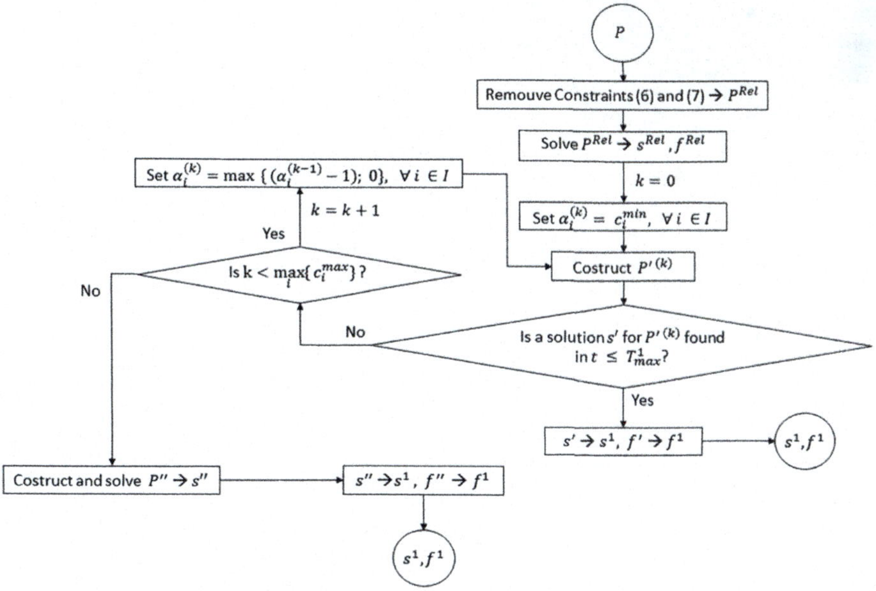

Fig. 1 Step 1: Identification of an initial feasible solution

4.2 Step 2: Improvement of the Initial Solution

The iterative methodology, aimed at improving the starting solution, uses a simple destruction operator to generate neighborhoods. In order to assess the general validity of the methodology, in this study we are focused only on a simple destruction operator that acts only on days. Specifically, it operates by destroying entirely for all nurses the assignments of a certain set of days, which is chosen randomly. This does not preclude the development of more complex destroying strategies in future studies. The preliminary experiments we carried out suggested a good balance between the effectiveness of the iterations and the speed of solution reconstruction. Several levels of balance were tested by destroying a percentage of the current solution ranging between 5 and 15% in each iteration. Based on these results, we destroyed about 10% of the current assignments. The exact percentage value varies from iteration to iteration because we have chosen to destroy an entire day selected even if this leads to slightly exceeding the set percentage value. The subsequent reconstruction is done by solving the subproblem P^{Curr}, which is obtained as a result of the destruction. If, after T^2_{max} the optimal solution or, at least, a feasible solution for the subproblem has not yet been found, the iteration is stopped without updating the current solution, and the procedure goes to the next iteration. Conversely, the best solution found becomes the new current solution, and the procedure is repeated until the maximum number of iterations, called K, is reached. When this occurs, the best feasible solution found s^H is returned. Figure 2 summarizes this procedure.

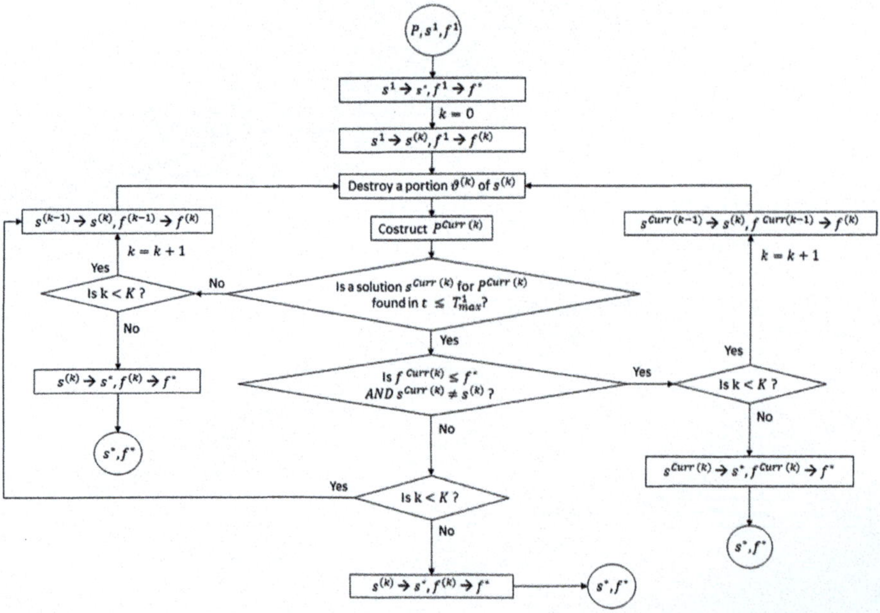

Fig. 2 Step 2: Iterative improvement of the solution

5 Computational Results

In this section, we solved some NRP instances of the Shift Scheduling Benchmark Dataset [5]. We summarize the specific instances addressed and then discuss the results obtained. This dataset contains 24 instances of varying size and difficulty: the smallest of these considers a planning horizon of 14 days, 8 nurses, and a single type of shift while the largest considers planning over 364 days, 150 nurses, and 32 different types of shifts. This is a preliminary study; therefore, we focused only on small and medium-sized instances. We solved the four instances ranging from Instance 5 to 8 and 10 of the dataset, whose main characteristics are described in Table 2. Note that all the instances chosen have significant sizes such that they are nontrivial problems and potentially representative of a real situation. The maximum run time was set equal to 10 min to make our experiments comparable with those in [3], which are, to the best of our knowledge, the best available solutions for 10-minute runs.

Table 2 Features of the addressed instances

Instance	Weeks	Nurses	Shift	Shift on request	Shift off request	Cover request
*I*_5	4	16	2	79	27	896
*I*_6	4	18	3	87	48	1512
*I*_7	4	20	3	104	64	1680
*I*_8	4	30	4	139	86	3360
*I*_10	4	40	5	210	74	5600

The proposed solution approach was implemented using the IBM ILOG CPLEX Optimization Studio package (Version 22.1.0), employing both OPL and IBM ILOG Script languages. Parameters T^1_{max} and T^2_{max} were both set to 100 s; parameter K was set to 6. The experimental phase was conducted on a machine having the following characteristics: DELL Optiplex 3040 SFF, Intel Core i5-6500 2.3 Ghz, 16 GB, 256 GB SSD, Nvidia GT 730 2GB, Windows 10 PRO.

The results and the computational times are shown in Table 3. They are compared with those obtained from an exact CPLEX-based approach, with the best results found in [3, 17], where the same run time limit equal to 10 min was used. In the last column there are the best-known solutions (BKS) reported in [5]. We have not compared our results with those found in [14, 16] because in these two papers there are not results on the same instances over 10-minute runs.

We observe that CPLEX always performs well because the instances addressed are not very large. Indeed, for four out five instances the optimality gap is 0%. In addition, we observe that our approach always terminates because of the maximum number of iterations (i.e. $K = 6$) was reached. This could explain the difference, in terms of a worst objective function value, for Instance I_5, I_6, I_7 and I_{10} with respect to the other results. In solving I_8, which appears to have a more complex combinatorial structure, FiNeMath reported the best results, managing to generate a

Table 3 Result and comparison. The best heuristic solutions found by the analyzed methods are in bold. The best known solutions are underlined. Times are expressed in MM:SS format

Instance	FiNeMath		CPLEX			MIP-DNN [3]		HA [17]		BKS [5]
	Value	Time	Value	Time	GAP (%)	Value	Time	Value	Time	Value
I_5	1151	00:30	**1143**	00:44	0.00	1146	10:00	**1143**	10:00	<u>1143</u>
I_6	1956	02:03	**1950**	00:31	0.00	**1950**	10:00	**1950**	10:00	<u>1950</u>
I_7	1062	00:55	**1056**	03:14	0.00	**1056**	10:00	**1056**	10:00	<u>1056</u>
I_8	**1328**	08:50	1457	10:00	12.05	1781	10:00	1364	10:00	<u>1300</u>
I_10	4632	02:34	**4631**	02:00	0.00	4632	10:00	**4631**	10:00	<u>4631</u>

solution close to the optimum in only 530 s. This is very promising for us, as it seems that our approach is able to generate good solutions for complex instances, where exact approaches fail. Further research is needed to verify this. Overall, however, the results show that our proposed procedure was able to determine very good solutions in a very short time for all the instances analyzed.

6 Conclusion and Future Work

In this paper, we addressed the NRP problem in the version proposed in the Shift Scheduling Benchmark Dataset [5]. We propose a matheuristics that operates in two steps: in the first, an initial feasible solution is constructed by passing through the optimal solution of a relaxed problem, and in the second, the current solution is improved through an approach combining F&O and LNS logics. A simple destruction operator that works by eliminating entire portions of the starting roster has been proposed for the construction of neighborhoods. The results obtained are promising because the approach was able to find good solutions in a reasonable time for all instances studied.

We suggest further research focusing both on designing useful methods to quickly determine a good initial solution and operators to generate neighborhoods that make the proposed heuristics even more efficient and effective so that these can also be used to cope with larger instances.

References

1. Burke, E., De Causmaecker, P., Vanden Berghe, G., Van Landeghem, H.: The state of the art of nurse rostering. J. Sched. **7**, 441–499 (2004)
2. Ceschia, S., Dang, N., De Causmaecker, P., Haspeslagh, S., Schaerf, A.: The second international nurse rostering competition. Ann. Oper. Res. **274**, 171–186 (2019)
3. Chen, Z., De Causmaecker, P., Dou, Y.: A combined mixed integer programming and deep neural network-assisted heuristics algorithm for the nurse rostering problem. Appl. Soft Comput. **136** (2023)

4. Chen, Z., Dou, Y., De Causmaecker, P.: Neural networked-assisted method for the nurse rostering problem. Comput. Indus. Eng. **171** (2022)
5. Curtois, T.: Employee shift scheduling benchmark data sets (2014). http://www.schedulingbenchmarks.org/. Accessed 20 Jan. 2024
6. Curtois, T., Qu, R.: Computational results on new staff scheduling benchmark instances. Technical Report, ASAP Research Group, School of Computer Science, University of Nottingham, NG8 1BB. Nottingham, UK (2014)
7. den Hartog, S.G.M., Hoogeveen, H., van der Zanden, T.C.: On the complexity of nurse rostering problems. Oper. Res. Lett. **51**(5), 483–487 (2023)
8. Gander, P., O'Keeffe, K., Santos-Fernandez, E., Huntington, A., Walker, L., Willis, J. : Development and evaluation of a matrix for assessing fatigue-related risk, derived from a national survey of nurses' work patterns. Int. J. Nurs. Stud. 112 (2020)
9. Gintner, V., Kliewer, N., Suhl, L.: Solving large multiple-depot multiple-vehicle-type bus scheduling problems in practice. OR Spectrum **27**, 507–523 (2005)
10. Guido, R., Groccia, M.C., Conforti, D.: An efficient matheuristic for offline patient-to-bed assignment problems. Eur. J. Oper. Res. **268**(2), 486–503 (2018)
11. Haspeslagh, S., De Causmaecker, P., Schaerf, A., Stolevik, M.: The first international nurse rostering competition 2010. Ann. Oper. Res. **218**, 221–236 (2014)
12. Ngoo, C.M., Goh, S.L., Sze, S.N., Sabar, N.R., Abdullah, S., Kendall, G.: A survey of the nurse rostering solution methodologies: the state-of-the-art and emerging trends. IEEE Access **10**, 56504–56524 (2022)
13. Shaw, P.: Using constraint programming and local search methods to solve vehicle routing problems. In: Maher, M., Puget, JF. (eds.) Principles and Practice of Constraint Programming. Lecture Notes in Computer Science, vol. 1520. Springer, Berlin, Heidelberg. (1998)
14. Smet, P.: Constraint reformulation for nurse rostering problems. In: Proceedings of the 12th International Conference of the Practice and Theory of Automated Timetabling, pp. 69–80 (2018)
15. Smet, P., Brucker, P., De Causmaecker, P., Vanden Berghe, G.: Polynomially solvable personnel rostering problems. Eur. J. Oper. Res. **249**(1), 67–75 (2016)
16. Strandmark, P., Qu, Y., Curtois, T.: First-order linear programming in a column generation-based heuristic approach to the nurse rostering problem. Comput. Oper. Res. (2020)
17. Rahimian, E., Akartunalı, K., Levine, J.: A hybrid integer programming and variable neighbourhood search algorithm to solve nurse rostering problems. Eur. J. Oper. Res. **258**(2), 411–423 (2017)
18. Römer, M., Mellouli, T.: A direct MILP approach based on state-expanded network flows and anticipation for multi-stage nurse rostering under uncertainty. In: Proceedings of the 11th International Conference on the Practice and Theory of Automated Timetabling, pp. 549–551 (2016)
19. Van den Bergh, J., Beliën, J., De Bruecker, P., Demeulemeester, E., De Boeck, L.: Personnel scheduling: a literature review. Eur. J. Oper. Res. **226**(3), 367–385 (2013)

Mixed-Integer Linear Programming Models for Water-Energy-Food-Land Nexus Optimization with Land Degradation Phenomenon

Chiara Maragó

Abstract The Water-Energy-Food Nexus planning is grounded on the three pillars of sustainability: economy, society, and environment. Addressing the needs of the steadily growing global population has emerged as an increasingly urgent challenge. The inherent conflicts among objectives within economic, social, and environmental dimensions and the shortage of required resources are exacerbating the pressure. Land degradation, together with climate change and other disruptive phenomena, is making difficult to meet the growing demand for food. Thus, conscious decision-making processes on all the three decision-making levels, i.e., strategic, tactical, and operational, require quantitative analyses which take into account the current situation. As a novel contribution, in this work two Single-Objective Mixed-Integer Linear Programming models based on a Water-Energy-Food-Land Nexus system are proposed. Then, a what-if analysis based on hypothetical real-world scenarios, considering reductions in crop yields and changes in fossil fuel availability, is carried out.

Keywords Mixed-integer linear programming · Land degradation · Water-energy-food-land nexus · Security assessment

1 Introduction

In recent years, meeting the growing demand for food as a result of population growth has been the main challenge for the agricultural sector. At the same time, the mounting pressure on using scarce resources such as water, energy, and land is encouraging significant changes towards more sustainable production and consumption patterns. Hoff [3] stated that agricultural production will have to increase by about 70% by 2050, whereas agricultural land will have to expand by about 10% globally, for food

C. Maragó (✉)
Department of Mechanical, Energy and Management Engineering, University of Calabria, Arcavacata, Italy
e-mail: chiara.marago@unical.it; maragochiara@gmail.com

M. Barbato et al. (eds.), *Beyond Frontiers of Operations Research: Emerging Technologies and Innovative Optimization Paradigms*, AIRO Springer Series 14,
https://doi.org/10.1007/978-3-031-85894-9_9

production and bio-fuels generation. The main problems in such a context are the shortage of available land and the land degradation (LD), which is one of the main consequences of intensive agricultural production combined with the use of chemical fuels. By 2050, it is estimated that this phenomenon, in conjunction with climate change, will result in a reduction in global crop yields of approximately 10% globally and up to 50% in some regions [4]. Furthermore, it was observed that agriculture is responsible for a significant proportion of fresh water over-exploitation, consuming an average of 70% of fresh water and up to 90% in some countries. The Water-Energy-Food Nexus (WEF Nexus) approach highlights the interdependence of water, energy, and food sectors and the synergies among them [3]. The evidence-based need for sustainable WEF Nexus-based plans is pushing the Operations Research community to develop efficient and innovative decision support tools at different levels of decision-making, i.e., strategic, tactical, and operational. In particular, decisions at the strategic level have political and social implications on a large scale, are uncertain, and focused on the overall purpose of a project [6]. In the literature, different methodologies have been proposed at this level: the augmented ϵ-constraint method was applied to solve a multi-period (MP) multi-objective (MO) stochastic mixed-integer linear programming (MILP) model [7], aiming to design a sustainable supply chain under multiple uncertainties; simulation-optimization approaches were proposed to identify the most suitable pathways for utilising Jatropha curcas plant as bio-fuel feedstock [1] and to explore long-term alternative cropping strategies while considering an acceptable maximum reduction ratio of agricultural land [8]. Some objective functions in economic, environmental, social, and productivity dimensions are: minimizing total costs, resource consumption and pollutant emissions, maximizing social acceptance of decisions and produced output. To guarantee the security of resources while evaluating economic and environmental aspects, a bi-level (BL) MP and a MP optimization model were proposed in [10, 11], respectively. These models were developed to meet socioeconomic demands while considering the inner requirements of the nexus system, the limited availability of resources, and the environmental impacts resulting from resource use. A MO MILP model useful to assess real-world WEF Nexus scenarios was developed in [5]. As a novel contribution to the literature, two MILP models, based on [10, 11], are formulated to perform a MP what-if analysis based on hypothetical time-varying real-world scenarios, i.e., reductions in crop yields due to LD and changes in fossil fuel availability for recent green policies. The aim is to provide a basis for discussion of the economic and environmental impact of these changes on the water, energy, food and land nexus (WEFL Nexus) resources and insights for their strategic management. The rest of the paper is organized as follows. Section 2 presents the problem under study along with the MILP models. Section 3 illustrates the WEFL Nexus scenarios considered as case studies and then discusses the results.

2 Problem Statement and Model Formulation

In the BL MILP model proposed in [10], the considered elements of a nexus system are: electricity generated by fossil fuel (i.e., coal and natural gas) power plants (PPs) with limited capacity and capacity expansion (CE) options; water distinguished as surface, ground, and recycled water. Decisions concern conflicting objectives. At the upper level the optimal allocation of resources aims to maximize the total generated electricity from the available PPs, while at the lower level the total system cost is minimized. The overall planning horizon is partitioned into smaller and consecutive planning periods. The optimal allocation of the WE Nexus resources in each time period is constrained by limited existing capacity of the PPs; their capacity optionally could be increased at a given economic cost. Food sector was added to the previous model in [11] and the goal was to minimize the total cost function (*TC*) of a WEF Nexus system, given as the sum of four components: F_1 is the overall energy cost computed as sum of fossil fuels supply cost, electricity generation cost, and cost of expansion of PPs; F_2 is the overall water consumption cost, i.e., the sum of water consumption costs for food production and electricity generation; F_3 is the food production cost; F_4 is the cost of abatement of CO_2 emissions, which are attributable to both the food and energy sectors. The WEF Nexus system constraints can be grouped as follows:

- fossil fuels availability; water availability for food production and electricity generation; electricity availability for food production and for water collection, treatment and delivery;
- electricity and food demands satisfaction; water demand satisfaction for food production and for electricity generation;
- limitation of greenhouse gas emissions.

Although the food sector was then considered, neither the harvested area nor the impact of the LD phenomenon on crop yields was assessed. In the following, two MILP models are formulated, considering harvested area and crop yields, reduced by LD, for food production. They are based on those of [10, 11]. Specifically, both the proposed models support a decision-maker in managing a WEFL Nexus system with different scenarios of increased demands, considering CE through capacity modification of existing facilities (Sect. 2.1) or construction of new PPs (Sect. 2.2). Although it is possible to add a new plant and increase the capacity of existing ones at the same time, our aim was to assess the impact of the two different decisions taken independently, under hypothetical scenarios, both in economic and environmental terms. The sets, indices, parameters and decision variables of the proposed models are listed in Table 1.

Table 1 Sets, parameters, and decision variables

Sets	
J	Set of l existing power plants and fossil fuels, indexed by $j = 1, \ldots, l$
$\bar{J}$	Set of p new power plants, indexed by $j = 1, \ldots, p$
I	Set of q fossil fuels, indexed by $i = 1, \ldots, q$
T	Set of k planning periods, indexed by $t = 1, \ldots, k$
M	Set of s expansion options, indexed by $m = 1, \ldots, s$
Parameters	
α_j	Unit water demand per unit of electricity generation in the power plant j
E^{max}	Maximum allowable CO_2 emissions during the overall planning period
μ_j, δ	Loss factor of water delivery to the power plant j and the food sector
$\bar{C}_j$	Fixed costs for electricity generation for the power plant j
$\bar{E}_j$	Existing capacity for the power plant j
C_{jt}, C_{it}	Average costs of fossil fuel j or i supply in period t
C_{jt}^{eg}	Electricity generation average costs in power plant j in period t
E_{jt}^{eg}	Unit electricity production per unit of capacity in power plant j in period t
C_{jt}^{ex}	Costs for capacity expansion for the power plant j in period t
c_{jmt}^{ex}	Expanded capacity of power plant j with option m
$C_{jt}^{gw(e)}, C_{jt}^{sw(e)}, C_{jt}^{rw(e)}$	Supply costs of ground, surface and recycled water for the power plant j in period t
$C_t^{gw(f)}, C_t^{sw(f)}$	Supply costs of groundwater and surface water for food production in period t
C_t^f	Unit costs of food production
$C_t^{a(e)}, C_t^{a(f)}$	CO_2 abatement costs for electricity generation and food production in period t
O_t^f, O_{jt}^e	Unit of CO_2 emissions per unit of food production and per unit of electricity generation in power plant j in period t
E_{jt}	Unit of energy carrier per unit of electricity generation for power plant j in period t
E_t^f, E_t^w	Unit energy demand for food production and water sector in period t
W_t^f	Unit water consumption per unit of food production in period t
$b_t^{e(f)}, b_t^{e(w)}$	Maximum available electricity for food production and water sector in period t
$b_t^{gw}, b_t^{sw}, b_t^{rw}$	Maximum available ground, surface, and recycled water amount in period t

(continued)

Table 1 (continued)

b_{jt}, b_{it}	Availability of fossil fuel j or i in period t
D_t^e, D_t^f	Socioeconomic demands of electricity and food in period t
DC_t	Land degradation coefficient in period t
Y_t	Crop yield in period t
ϕ_{jt}	CO_2 abatement average efficiency in power plant j in period t
Decision variables	
$x_{jt}, x_{it} \geq 0$	Fossil fuel j or i supply in period t
$x_{jt}^{eg}, k_{jt}^{eg} \geq 0$	Electricity generation from power plant j in J and $\bar{J}$, respectively, in period t
$w_t^{g(f)}, w_t^{s(f)} \geq 0$	Ground and surface water supplied to food sector in period t
$w_{jt}^{g(e)}, w_{jt}^{s(e)}, w_{jt}^{r(e)} \geq 0$	Ground, surface, and recycled water supplied in period t to power plant $j \in J$
$\bar{w}_{jt}^{g(e)}, \bar{w}_{jt}^{s(e)}, \bar{w}_{jt}^{r(e)} \geq 0$	Ground, surface, and recycled water supplied in period t to power plant $j \in \bar{J}$
$A_t \geq 0$	Area harvested in period t
$z_{jmt} \in \{0, 1\}$	1 if power plant j is expanded with option m in period t, 0 otherwise
$y_j \in \{0, 1\}$	1 if power plant j in $\bar{J}$ is activated, 0 otherwise

2.1 A MILP Model for a WEFL Nexus System with Capacity Expansion Options for Existing Power Plants

The first MILP model is formulated as follows.

$$TC = \min F_1 + F_2 + F_3 + F_4 \tag{1}$$

$$F_1 = \sum_{j \in J} \sum_{t \in T} C_{jt} x_{jt} + \sum_{j \in J} \bar{C}_j + \sum_{j \in J} \sum_{t \in T} C_{jt}^{eg} x_{jt}^{eg} + \sum_{j \in J} \sum_{m \in M} \sum_{t \in T} C_{jt}^{ex} c_{jmt}^{ex} z_{jmt} \tag{2}$$

$$F_2 = \sum_{t \in T} (C_t^{gw(f)} w_t^{g(f)} + C_t^{sw(f)} w_t^{s(f)}) + \sum_{j \in J} \sum_{t \in T} (C_{jt}^{gw(e)} w_{jt}^{g(e)} + C_{jt}^{sw(e)} w_{jt}^{s(e)} + C_{jt}^{rw(e)} w_{jt}^{r(e)}) \tag{3}$$

$$F_3 = \sum_{t \in T} C_t^f Y_t A_t \tag{4}$$

$$F_4 = \sum_{j \in J} \sum_{t \in T} O^e_{jt} C^{a(e)}_t x^{eg}_{jt} + \sum_{t \in T} O^f_t C^{a(f)}_t Y_t A_t \tag{5}$$

$$E_{jt} x^{eg}_{jt} \le x_{jt} \qquad \forall j, t \tag{6}$$

$$x^{eg}_{jt} \le E^{eg}_{jt} (\bar{E}_j + \sum_{m \in M} \sum_{t_1 = 1 \ldots t} c^{ex}_{jmt_1} z_{jmt_1}) \qquad \forall j, t \tag{7}$$

$$x_{jt} \le b_{jt} \qquad \forall j, t \tag{8}$$

$$E^f_t Y_t A_t \le b^{e(f)}_t \qquad \forall t \tag{9}$$

$$E^w_t (w^{g(f)}_t + w^{s(f)}_t + \sum_{j \in J} (w^{g(e)}_{jt} + w^{s(e)}_{jt} + w^{r(e)}_{jt})) \le b^{e(w)}_t \qquad \forall t \tag{10}$$

$$\sum_{j \in J} x^{eg}_{jt} - E^f_t Y_t A_t - E^w_t (w^{g(f)}_t + w^{s(f)}_t$$

$$+ \sum_{j \in J} (w^{g(e)}_{jt} + w^{s(e)}_{jt} + w^{r(e)}_{jt})) \ge D^e_t \qquad \forall t \tag{11}$$

$$(1 - \delta)(w^{g(f)}_t + w^{s(f)}_t) \ge W^f_t Y_t A_t \qquad \forall t \tag{12}$$

$$(1 - \mu_j)(w^{g(e)}_{jt} + w^{s(e)}_{jt} + w^{r(e)}_{jt}) \ge \alpha_j x^{eg}_{jt} \qquad \forall j, t \tag{13}$$

$$w^{g(f)}_t + \sum_{j \in J} w^{g(e)}_{jt} \le b^{gw}_t \qquad \forall t \tag{14}$$

$$w^{s(f)}_t + \sum_{j \in J} w^{s(e)}_{jt} \le b^{sw}_t \qquad \forall t \tag{15}$$

$$\sum_{j \in J} w^{r(e)}_{jt} \le b^{rw}_t \qquad \forall t \tag{16}$$

$$(1 - DC_t) Y_t A_t \ge D^f_t \qquad \forall t \tag{17}$$

$$\sum_{j \in J} \sum_{t \in T} x^{eg}_{jt} O^e_{jt} (1 - \phi_{jt}) + \sum_{t \in T} O^f_t Y_t A_t \le E^{max} \tag{18}$$

$$\sum_{m \in M} z_{jmt} \le 1 \qquad \forall j, t \tag{19}$$

$$x_{jt}, x^{eg}_{jt}, w^{g(e)}_{jt}, w^{s(e)}_{jt}, w^{r(e)}_{jt} \ge 0 \qquad \forall j, t \tag{20}$$

$$w^{g(f)}_t, w^{s(f)}_t, A_t \ge 0 \qquad \forall t \tag{21}$$

$$z_{jmt} \in \{0, 1\} \qquad \forall j, m, t \tag{22}$$

The objective function (1) is the minimization of TC on the overall planning horizon T. All the constraints, except Constraints (18), are formulated per each period $t \in T$ in which the planning horizon has been partitioned. Constraints (6) ensure that the generated electricity converted from supplied fossil fuels should not be greater than the amount of fossil fuel supply in each PP. Constraints (7) require that electricity generation in each PP is limited by the sum of existing and possible expanded capacity. Furthermore, they ensure that decisions on CE in one period remain valid for subsequent periods. Constraints (8) ensure that the fossil fuel supply is not greater than the availability for each resource. Constraints (9) and (10) guarantee that the energy demand for food production and water collection, treatment and delivery is not greater than the maximum available electricity for food and water sector, respectively. Constraints (11) ensure the electricity demand satisfaction while satisfying the electricity requirements from food and water sector. Constraints (12) and (13) impose the satisfaction of water demand for food production and electricity generation, taking into consideration a loss factor in water delivery for both sectors. Constraints (14)–(16) limit the water consumption in food and energy sectors to the maximum availability for each water source. Constraints (17) guarantee that the amount of food produced, given by the product of the cultivated area and the crop yield, at least meets the demand for food while considering a LD coefficient. Constraints (18) state that the generated CO_2 amounts from food and energy sectors should not be larger than the maximum allowable CO_2 emissions during the overall planning period. Constraints (19) ensure that the capacity of each PP can be expanded only once at the beginning of each planning period. Finally, Constraints (20)–(22) specify the domain of the variables. Hereafter, this MILP model will be referred to as Model 1.

2.2 A MILP Model for a WEFL Nexus with Capacity Expansion Through Activation of New Power Plants

The second model differs from Model 1 in the objective function, in some constraints, additional sets and variables. Indeed, we introduced the sets of PPs fired with fossil fuel $i \in I$ as $J_i \subseteq J$ and $\bar{J}_i \subseteq \bar{J}$. $\bar{J}$ is the set of new PPs. The set I of fossil fuels was introduced because operating PPs could be more than two, considering capacity expansion through the activation of new PPs. Some constraints and equations are here reported. The full mathematical formulation can be found in [9]. Equation (2) is replaced by Equation (23); Constraints (6) and (7) are replaced by Constraints (24)–(26), respectively; Constraints (11) and (18) are replaced by Constraints (27) and (28), respectively. Finally, Constraints (29) and (30) specify the domain of the variables. Hereafter, this MILP model will be referred to as Model 2.

$$F_1 = \sum_{i \in I} \sum_{t \in T} C_{it} x_{it} + \sum_{j \in J} \bar{C}_j + \sum_{j \in \bar{J}} \bar{C}_j y_j + \sum_{j \in J} \sum_{t \in T} C_{jt}^{eg} x_{jt}^{eg}$$
$$+ \sum_{j \in \bar{J}} \sum_{t \in T} C_{jt}^{eg} k_{jt}^{eg} + \sum_{j \in \bar{J}} C_j^{ex} \bar{E}_j y_j \quad (23)$$

$$E_{jt} x_{jt}^{eg} + E_{jt} k_{jt}^{eg} \leq x_{it} \qquad \forall i \in I, \forall j \in (J_i \cup \bar{J}_i), \forall t \quad (24)$$

$$x_{jt}^{eg} \leq E_{jt}^{eg} \bar{E}_j \qquad \forall j \in J, \forall t \quad (25)$$

$$k_{jt}^{eg} \leq E_{jt}^{eg} \bar{E}_j y_j \qquad \forall j \in \bar{J}, \forall t \tag{26}$$

$$\sum_{j \in J \cup \bar{J}} (x_{jt}^{eg} + k_{jt}^{eg}) - E_t^f Y_t A_t - E_t^w (w_t^{g(f)} + w_t^{s(f)} + \sum_{j \in J \cup \bar{J}} (w_{jt}^{g(e)} + w_{jt}^{s(e)} + w_{jt}^{r(e)} + \bar{w}_{jt}^{g(e)} + \bar{w}_{jt}^{s(e)} + \bar{w}_{jt}^{r(e)})) \geq D_t^e \qquad \forall t \tag{27}$$

$$\sum_{j \in J \cup \bar{J}} \sum_{t \in T} (x_{jt}^{eg} + k_{jt}^{eg}) O_{jt}^e (1 - \phi_{jt}) + \sum_{t \in T} O_t^f Y_t A_t \leq E^{max} \tag{28}$$

$$x_{it}, k_{jt}^{eg}, \bar{w}_{jt}^{g(e)}, \bar{w}_{jt}^{s(e)}, \bar{w}_{jt}^{r(e)} \geq 0 \qquad \forall i, j \in \bar{J}, t \tag{29}$$

$$y_j \in \{0, 1\} \qquad \forall j \in \bar{J} \tag{30}$$

Equation (23) includes the cost of fossil fuel supply, electricity generation costs (including fixed costs), and capital costs for activation of new PPs. Constraints (25) and (26) limit the quantity of electricity generated to the capacity of existing and new PPs in each planning period, respectively. Constraints (24), (27) and (28), as well as the remaining unwritten equations and constraints, add the contribution of new PPs in electricity generation and water consumption.

3 Case Study: Real-World WEFL Nexus Scenarios

To test the two optimization models as decision-making tools at strategic level, several computational experiments were carried out. The data used are real-world information taken from [2, 4, 10, 11]. The planning horizon has 15 years and it is divided into three 5-year consecutive periods. In [10, 11] only two PPs are considered (one coal-fired PP (CPP) and one natural gas-fired PP (NGPP)), and the possibility of CE. Table 2 lists data for each period and PP. It is assumed that:

- Expansion options: 0.26, 0.31, and 0.36 (GW) for CPPs; 0.09, 0.14, and 0.19 (GW) for NGPPs; they are constant upon the three periods.
- Additional PPs: one CCP (0.9 GW) and one NGPP (0.5 GW), with capital costs of 700 and 550 (mln \$/GW), respectively.
- Two parameters are constant for all indexes: $E^{max} = 19,000$ (mln kg), and $\delta = 15$ (%).

In 2018, it was estimated that the LD phenomenon will reduce crop yields by an average of 10% globally by 2050, and up to 50% in some regions [4]. Therefore, five different scenarios related to this percentage reduction coefficient were considered to

Table 2 Parameter values

	Time periods			Unit of measure		Time periods			Unit of measure
	$t = 1$	$t = 2$	$t = 3$			$t = 1$	$t = 2$	$t = 3$	
$C^{gw(f)}$	1.96	2.42	2.83	$/10^3 gal	D^f	67	71	75	10^3 t
$C^{sw(f)}$	2.23	2.56	3.38	$/10^3 gal	DC	1.67	3.34	5.01	%
C^f	149.5	165.5	180.0	$/t	D^e	142	155	163	PJ
$C^{a(e)}$	13.2	14.1	15.8	10^3 $/mln kg	W^f	659	676	694	10^3 gal/t
$C^{a(f)}$	10.8	11.9	13.1	$/t	b^{gw}	50	48	46	bln gal
$b^{e(f)}$	0.23	0.25	0.27	PJ	b^{sw}	32	30	27	bln gal
$b^{e(w)}$	1.15	1.2	1.25	PJ	b^{rw}	30	27	24	bln gal
E^f	2.52	2.64	2.75	PJ/10^6 t	Y_t	8.65	8.75	8.85	t/ha
E^w	3.56	3.79	3.91	KWh/10^3 gal	O^f	0.48	0.48	0.48	
	CPP					NGPP			
C	2.65	2.89	3.16	mln $/PJ		4.58	4.86	5.12	mln $/PJ
C^{eg}	0.20	0.22	0.24	mln $/PJ		0.46	0.49	0.52	mln $/PJ
E^{eg}	85	90	95	PJ/GW		75	80	85	PJ/GW
C^{ex}	700	660	620	mln $/ GW		550	500	450	mln $/ GW
$C^{gw(e)}$	1.78	2.03	2.47	$/10^3 gal		2.21	2.62	3.12	$/10^3 gal
$C^{sw(e)}$	2.06	2.55	2.93	$/10^3 gal		1.80	2.14	2.55	$/10^3 gal
$C^{rw(e)}$	3.89	4.05	4.31	$/10^3 gal		4.02	4.18	4.37	$/10^3 gal
O^e	261.03	254.89	247.08	mln kg/PJ		152.58	149.98	146.19	mln kg/PJ
E	3.1	2.9	2.7			2.5	2.3	2.1	
b	319	306	295	PJ		237	224	212	PJ
	CPP	NGPP				CPP	NGPP		
$\bar{C}$	55	65		mln $	μ	10	15		%
$\bar{E}$	0.9	0.5		GW	ϕ_t	80	85		%
α	91.74	122.32		10^6 gal/PJ					

perform a what-if analysis on a MP basis. The percentage reduction coefficient is 10% (s_1), 20% (s_2), 30% (s_3), 40% (s_4) and 50% (s_5), respectively, for each single scenario. As this estimate dates back to 2018, we considered the 30-year period between 2020 and 2050 as the reference period for the reduction coefficient. Furthermore, we assumed a constant reduction in crop yields over time. This value was then divided equally between two 15-year periods and further divided between three five-year time periods (TPs). For example, the average reduction in crop yields of 10% is considered to be about 5% over 15 years and about 1.67% for each TP. For the second and third TP, this percentage was calculated as a cumulative value. For our analysis, we used real-world crop yield data provided by [2]. We observed an average increase of about 1.12% in crop yields over 5-year time windows between 2000 and 2020, driven by the adoption of innovative techniques and pesticides. The year 2020 was taken as the base year. Its average crop yield value was estimated as the mean of the crop yield values for period 2019–2021. We applied then the observed average increase to estimate crop yield values for the subsequent three five-year TPs. We stress that these are not meant as detailed projections, but rather as a basis for discussion. In this study, we refer to the fossil fuel availability considered in [10] as a_0. The urgency of reducing the amount of coal consumed in favour of natural gas, led us to consider a different scenario, designated as a_1. In a_1, it was assumed that: the availability of coal is 7, 10, 13% lower than the values of a_0 for the first, the second, and the third TP, respectively. At the same time, it was assumed that the availability of natural gas increases of 10 , 13, 15% for each TP. The two scenarios a_0 and a_1 were then combined with each of the LD scenarios. The use of the two MILP models aims to provide insights into the economic and environmental dimensions. These models focus on the design of the optimal mix of facilities in the energy sector, i.e., deciding to expand existing PPs or add new ones, to meet food and electricity demands despite the LD.

3.1 Results and Discussion

The optimization models were implemented in OPL and solved with ILOG CPLEX optimization Studio IBM, version 22.1.1 with Academic License. Detailed computational results can be found in [9]. Some of them are reported here in Table 3. The results refer to the scenarios s_0 (the base scenario without LD), s_1, and s_2, in combination with scenarios a_0 and a_1. The results for scenarios s_3, s_4, and s_5 are not shown, as the models become infeasible due to three key parameters: electricity availability for the water sector, water sources availability, and electricity availability for the food sector, respectively. The existing power and water resources would then be insufficient to cope with the increasing use of land resources over time. The first part of Table 3 shows the values of area harvested and water consumed for each TP, considering Models 1 and 2. The values presented in brackets refer to scenario a_1. Each reported value was compared to the results under s_0 by a percentage value, written and shown in the columns having the heading Increase. When scenarios s_0 and s_1 occur, the area harvested do not vary depending on scenarios a_0 and a_1 for both models, and the water consumed under a_0 and a_1 for Model 2 is the same. Indeed, for Model 2, the new PPs activated to meet electricity demands are NGPPs, having lower capital costs. The greater availability of coal and capacity of CPPs and the lower operational costs of CPPs in both scenarios a_0 and a_1, leave the water consumption and the energy mix unchanged. Moreover, when a_1 occurs, both models become infeasible even under s_2, due to the higher contribution of NGPPs in the energy mix, requiring more water and then more electricity. The second part of Table 3 shows the total cost and the amount of CO_2 emissions of both models over the entire planning horizon. Only the results related to a_0 are here reported, for two main reasons: we found them more significant in terms of variations in costs and CO_2 emissions between models; the infeasibility of models under s_2 combined with a_1, as explained before. The percentages, reported in brackets, provide insights into the results. The activation of a novel PP would cost about 2.062, 2.065, and 10.44% more than expanding existing PPs, under s_0, s_1, and s_2, respectively. Furthermore, CO_2 emissions would be about 8.40% less under s_0 and s_1, because only NGPPs, less polluting, would be activated. Activating two additional PPs under s_2,

Table 3 Results of Models 1 and 2 under the combined scenarios

	First 5-year period		Second 5-year period		Third 5-year period	
Models 1 and 2: Area harvested (ha)						
$s_0 - a_0(a_1)$	7742 (7742)	**Increase**	8113 (8113)	**Increase**	8474 (8474)	**Increase**
$s_1 - a_0(a_1)$	7874 (7874)	1.70% (1.70%)	8393 (8393)	3.46% (3.46%)	8921 (8921)	5.27% (5.27%)
$s_2 - a_0(a_1)$	8009 (infeasible)	3.45% (–)	8692 (infeasible)	7.14% (–)	9416 (infeasible)	11.11% (–)
Model 1: Water consumed (bln gal)						
$s_0 - a_0(a_1)$	68.64 (68.52)	**Increase**	74.71 (74.96)	**Increase**	80.45 (80.90)	**Increase**
$s_1 - a_0(a_1)$	69.53 (69.40)	1.31% (1.29%)	76.66 (76.92)	2.62% (2.61%)	83.69 (84.14)	4.02% (4.00%)
$s_2 - a_0(a_1)$	70.46 (infeasible)	2.65% (–)	78.75 (infeasible)	5.41% (–)	87.27 (infeasible)	8.47% (–)
Model 2: Water consumed (bln gal)						
$s_0 - a_0(a_1)$	69.32 (69.32)	**Increase**	75.55 (75.55)	**Increase**	81.30 (81.30)	**Increase**
$s_1 - a_0(a_1)$	70.21 (70.21)	1.28% (1.28%)	77.50 (77.50)	2.59% (2.59%)	84.54 (84.54)	3.98% (3.98%)
$s_2 - a_0(a_1)$	70.01 (infeasible)	0.99% (–)	78.81 (infeasible)	4.31% (–)	87.68 (infeasible)	7.85% (–)
Models 1 and 2: Total cost(TC) (mln \$) and CO_2 emissions (mln kg)						
	$s_0 - a_0$		$s_1 - a_0$		$s_2 - a_0$	
	Model 1	Model 2	Model 1	Model 2	Model 1	Model 2
TC	6930	7072 (+2.062%)	6955	7099 (+2.065%)	6986	7715 (+10.44%)
CO_2	19,000	17,396 (−8.44%)	19,000	17,402 (−8.41%)	19,000	19,000 (–)

one NGPP and one CPP, would be more expensive than expanding existing PPs, and would pollute just as much. The results reported in this table then highlight the conflicting nature of economic and environmental objectives of this problem. For example, when an additional NGPP is activated, the system cost is higher and the environmental cost decreases, because of the nature of these PPs. Thus, decision is on behalf of decision-makers, who should carefully evaluate which dimension of sustainability to prioritize. Stochastic programming will be one of future directions for other developments of this work.

References

1. Alherbawi, M., AlNouss, A., McKay, G., Al-Ansari, T.: Optimum sustainable utilisation of the whole fruit of *Jatropha curcas*: an energy, water and food nexus approach. Renew. Sustain. Energy Rev. **137**, 110605 (2021)
2. FAO. https://www.fao.org/faostat/en/#data/QCL. Accessed 15 June 2024
3. Hoff, H.: Understanding the Nexus. Background Paper for the Bonn2011. The Water, Energy and Food Security Nexus. Stockholm Environment Institute, Stockholm (2011)
4. IPBES: The IPBES assessment report on land degradation and restoration. In: Montanarella, L., Scholes, R., Brainich, A. (eds.). Secretariat of the Intergovernmental Science-Policy Platform on Biodiversity and Ecosystem Services, Bonn, Germany, 744 p. (2018)
5. Maragó, C., Guido, R., Guerriero, F.: Multi-objective optimization for the security of water-energy-food nexus. In: Bruglieri, M., Festa, P., Macrina, G., Pisacane, O. (eds.) Optimization in Green Sustainability and Ecological Transition. ODS 2023. AIRO Springer Series, vol. 12. Springer, Cham (2024). https://doi.org/10.1007/978-3-031-47686-0_26
6. Merad, M., Verdel, T., Roy, B., Kouniali, S.: Contribution of the multi-criteria decision aid methods for natural risk analysis and risk management studies. In: 2023 SRA Annual Meeting (2003)
7. Osmani, A., Zhang, J.: Multi-period stochastic optimization of a sustainable multi-feedstock second generation bioethanol supply chain—a logistic case study in Midwestern United States. Land Use Policy **61**, 420–450 (2017)
8. Yin, Z., Wu, J., Song, J., Yang, Y., Zhu, X., Wu, J.: Multi-objective optimization-based reactive nitrogen transport modeling for the water-environment-agriculture nexus in a basin-scale coastal aquifer. Water Res. **212**, 118111 (2022)
9. Zenodo. https://t.ly/Kqzoy. Accessed 9 Aug. 2024
10. Zhang, X., Vesselinov, V.V.: Energy-water nexus: balancing the tradeoffs between two-level decision makers. Appl. Energy **183**, 77–87 (2016)
11. Zhang, X., Vesselinov, V.V.: Integrated modeling approach for optimal management of water, energy and food security nexus. Adv. Water Resour. **101**, 1–10 (2017)

Author Index

M. Barbato et al. (eds.), *Beyond Frontiers of Operations Research: Emerging Technologies and Innovative Optimization Paradigms*, AIRO Springer Series 14,
https://doi.org/10.1007/978-3-031-85894-9

The manufacturer's authorised representative in the EU is Springer Nature Customer Service Centre GmbH, Europaplatz 3, 69115 Heidelberg, Germany. If you have any concerns regarding our products, please contact ProductSafety@springernature.com

Printed and bound by CPI Group (UK) Ltd, Croydon, CR0 4YY
30/09/2026
02225469-0003